GNU Gama Reference Manual

A catalogue record for this book is available from the Hong Kong Public Libraries.

Published in Hong Kong by Samurai Media Limited.

Email: info@samuraimedia.org

ISBN 978-988-8381-60-9

Table of Contents

1 Introduction

GNU Gama is a project dedicated to adjustment of geodetic networks. It is intended for use with traditional geodetic surveyings which are still used and needed in special measurements (e.g., underground or high precision engineering measurements) where the Global Positioning System (GPS) cannot be used.

In general, surveying is the technique and science of accurately determining the terrestrial or three-dimensional spatial position of points and the distances and angles between them.[1]

Adjustment is a technical term traditionally used by geodesists and surveyors which simply means "application of the least squares method to process the over-determined system of measurements" (statistical methods other than least squares are used sometimes but are not common). In other words, we have more observations than needed and we are trying to get the best estimate for adjusted observations and/or coordinates.

Adjustment of geodetic networks means that we have a set of points with given coordinates coordinates of some points and a set of observations among them. What is typical of adjustment of special geodetic measurements is that the resulting linearised system might be singular (we can have a network with no fixed points) and we are not only interested in the values of 'adjusted parameters and observations' but also in the estimates of their covariances. This is what Gama does.

Gama was originally inspired by Fortran system Geodet/PC (1990) designed by Frantisek Charamza. The GNU Gama project started at the department of mapping and cartography, faculty of Civil Engineering, Czech Technical University in Prague (CTU) about 1998 and its name is an acronym for *geodesy and mapping*. It was presented to a wider public for the first time at FIG Working Week 2000 in Prague and then at FIG Workshop and Seminar at HUT Helsinki in 2001.

The GNU Gama home page is
                    http://www.gnu.org/software/gama/
and the project is hosted on
                    http://savannah.gnu.org/git/?group=gama

GNU Gama is released under the GNU General Public License and is based on a C++ library of geodetic classes and functions and a small C++ template matrix library `matvec`. For parsing XML documents GNU Gama calls the `expat` parser version 1.1, written by James Clark. The `expat` parser is not part of the GNU Gama project, and is simply used by GNU Gama.

Adjustment in local Cartesian coordinate systems is fully supported by a command-line program `gama-local` that adjusts geodetic (free) networks of observed distances, directions, angles, height differences, 3D vectors and observed coordinates (coordinates with given variance-covariance matrix). Adjustment in global coordinate systems is supported only partly as a `gama-g3` program.

---

[1] Wikipedia, http://en.wikipedia.org/wiki/Surveying

1.1 Download

GNU Gama can be found in the subdirectory **/gnu/gama/** on your favourite GNU mirror or checked-out from the GIT. See our project page at savannah for more information.

To get an anonymous read-only access to the GIT repository for the latest GNU Gama source, issue the following command

```
git clone git://git.sv.gnu.org/gama.git
```

The collection of sample networks is available separetely. To checkout the **gama-local** examples from GIT use the command

```
git clone git://git.sv.gnu.org/gama/examples.git
```

1.2 Install

GNU Gama is developed and tested under GNU/Linux. A static library `libgama.lib` and executables are build in folders `lib` and `bin`. You can compile Gama easily yourself if you download the sources. If **expat** XML parser is installed on your system, change to the directory of Gama project and issue the following commands at the shell prompt

```
$ ./configure
$ make
```

To run tests from the Gama test suite try

```
$ make check
```

If the script **configure** is not available (which is the case when you download source codes from a git server), you have to generate it using auxiliary script **autogen.sh**. To compile and build all binaries. Run

```
$ ./configure [--bindir=DIR --infodir=DIR]
$ make install
```

if you want also to install the binaries. You can use configure parameters if you need to change directories where user executables and info documentation should be installed.

Typically, if you want to download (see Section 1.1 [Download], page 2) and compile sources, you will run following commands:

```
$ git clone git://git.sv.gnu.org/gama.git gama
$ cd gama
$ ./autogen.sh
$ ./configure
$ make
```

You should have **expat** XML parser and SQLite library already installed on your system. For example to be able to compile Gama on Ubuntu 10.04 you have to install following packages:

```
make doxygen git automake autoconf libexpat1-dev libsqlite3-dev
```

To compile user documentation in various formats (PDF, HTML, ...) run the following commands (before you have to run at least ./configure).

```
$ cd doc/
$ make download-gendocs.sh
$ make run-gendocs.sh
```

The documentation should be in `doc/manual` directory. To compile API documentation run

```
$ doxygen
```

in your `gama` directory. Doxygen output will be in the `doxygen` directory.

1.2.1 Precompiled binaries for Windows

GNU Gama builds for Windows are available from
https://sourceforge.net/projects/gnu-gama-builds/

1.3 Program `gama-local`

Program `gama-local` is a simple command line tool for adjustment of geodetic *free networks*. It is available for GNU Linux (the main platform on which project GNU Gama is being developed) or for MS Windows (tested with Borland compiler from Borland free command line tools and with Microsoft Visual C++ compiler; support for Windows platform is currently limited to maintaing compatibility with the two mentioned compilers).

Program `gama-local` reads input data in XML format (Chapter 2 [XML input data format for gama-local], page 7) and prints adjustment results into ASCII text file. If output file name is not given, input file name with extension `.txt` is used. If development files for Sqlite3 (package `libsqlite3-dev`) are installed during the build, `gama-local` also supports reading adjustment input data from an sqlite3 database. If run without arguments `gama-local` prints a short help

```
$ ./gama-local

Adjustment of local geodetic network            version: 1.16 / GNU g++
************************************
http://www.gnu.org/software/gama/


Usage: gama-local   input.xml  [options]
       gama-local   input.xml  --sqlitedb sqlite.db  --configuration name  [options]
       gama-local   --sqlitedb sqlite.db  --configuration name  [options]
       gama-local   --sqlitedb sqlite.db  --readonly-configuration name  [options]

Options:

--algorithm  svd | gso | cholesky | envelope
--language   en | ca | cz | du | es | fi | fr | hu | ru | ua | zh
--encoding   utf-8 | iso-8859-2 | iso-8859-2-flat | cp-1250 | cp-1251
--angles     400 | 360
--latitude   <latitude>
--ellipsoid  <ellipsoid name>
--text       adjustment_results.txt
--html       adjustment_results.html
--xml        adjustment_results.xml
--svg        network_configuration.svg
--cov-band   covariance matrix of adjusted parameters in XML output
```

```
              n  =  -1   for full covariance matrix (implicit value)
              n  >=   0   covariances are computed only for bandwidth n
    --iterations maximum number of iterations allowed in the linearized
              least squares algorithm (implicit value is 5)
    --version
    --help
```

Program `gama-local` version is followed by information on compiler used to build the program (apart from GNU `g++` compiler, two other possibilities are `bcc` and `msc` for Borland and Microsoft compilers respectively, when build under Microsoft Windows).

Option `--algorithm` enables to select numerical method used for solution of the adjustment. Implicitly is used Singular Value Decomposition (`svd`), alternatively user can decide for block matrix algorithm GSO by Frantisek Charamza, based on Gram-Schmidt orthogonalization. In both these cases, project equations are solved directly without forming *normal equations*. Third possibility is to select Cholesky decomposition of semidefinite matrix of normal equations (`cholesky`).

Option `--language` selects language used in output protocol. For example, if run with option `--language cz`, `gama-local` prints output results in Czech languague using UTF-8 encoding. Implicit value is `en` for output in English.

Option `--encoding` enables to change inplicit UTF-8 output encoding to iso-8859-2 (latin-2), iso-8859-2-flat (latin-2 without diacritics), cp-1250 (MS-EE encoding) cp-12251 (Russian encoding).

Option `--angles` selects angular units to be used in output.

Options `--latitude` and/or `--ellipsoid` are used when observed vertical and/or zenith angles need to be transformed into the projection plane. If none of these two options is explicitly used, no corrections are added to horizontal and/or zenith angles. If only one of these options is used, then implicit value for `--latitude` is 45 degrees (50 gons) and implicit ellipsoid is WGS84. Mathematical formulas for the corrections is given in the following section.

Adjustment results (`--text` and `--xml`) can be redirected to standard output if instead of a file name is used "-" string. If no output is given, XML adjustment format is implicitly send to standard output.

Option `--cov-band` is used to reduce the number of computed covariances (cofactors) in XML adjustment output. Implicitly full matrix is written to XML output, which could degrade time efficiency for the `envelope` algorithm for sparse matrix solution. Explicit option for full covariance matrix is `--cov-band -1`, option `--cov-band 0` means that only a diagonal of covariance matrix is written to XML output, `--cov-band 1` results in computing the main diagonal and first codiagonal etc. If higher rank is specified then available, it is reduced do maximum possible value `dim-1`.

Option `--iterations` enables to set maximum number of iterations allowed in the linearized least squares algorithm. After the adjustment `gama-local` computes differences between adjusted observations computed from residuals and from adjusted coordinates. If the positional difference is higher than 0.5mm, approximate coordinates of adjusted points are updated and the whole adjustment is repeated in a new iteration. Implicit number of iterations is 5.

1.3.1 Reductions of horizontal and zenith angles

For evaluating of reductions of horizontal and zenith angles, `gama-local` computes a helper point P_1 in the center of the network. Horizontal and zenith angles observed at point P_2 are transformed to the projection plane perpendicular to the normal z_1 of the helper point P_1. Coordinates (x_2, y_2) of point P_2 are conserved, but its normal z_2 is rotated by the central angle $2\gamma_{12}$ to be parallel with z_1.

For observations from point P_2 to point P_3 we denote the zenith angle z_{23}^m and horizontal direction σ_{23}^m. Now, transformed zenith angle z_{23} and horizontal direction σ_{23} can be expressed as

$$\cos z_{23} = \cos z_{23}^m \cos 2\gamma_{12} + \sin z_{23}^m \cos(180° - \sigma_{23}^m) \sin \gamma_{12},$$

$$\sin(180° - \sigma_{23}^m) \cot \sigma_{23} = -\cos(180° - \sigma_{23}^m) \cos 2\gamma_{12} + \cot z_{23}^m \sin 2\gamma_{12}$$

and after arrangement

$$\cos z_{23} = \cos z_{23}^m \cos 2\gamma_{12} - \sin z_{23}^m \cos \sigma_{23}^m \sin \gamma_{12},$$

$$\cot \sigma_{23} = \cot \sigma_{23}^m \cos 2\gamma_{12} + \frac{\cot z_{23}^m \sin 2\gamma_{12}}{\sin \sigma_{23}^m}$$

These formulas can be simplified for small networks, roughly up to the size of 6 kilometers, where

$$\cos 2\gamma_{12} \approx 1 \qquad \text{and} \qquad \sin 2\gamma_{12} \approx \frac{2\gamma_{12}''}{\varrho''}.$$

and

$$\cos z_{23} = \cos z_{23}^m - \sin z_{23}^m \cos \sigma_{23}^m \frac{2\gamma_{12}''}{\varrho''},$$

$$\cot \sigma_{23} = \cot \sigma_{23}^m + \frac{1}{\sin^2 \sigma_{23}^m} \cot z_{23}^m \sin \sigma_{23}^m \frac{2\gamma_{12}''}{\varrho''}.$$

Comparing these expressions with first members of Taylor series

$$f(x) \approx f(x^0) + \frac{\mathrm{d}f(x^0)}{\mathrm{d}x}$$

of functions $\cos z_{23}$ and $\cot \sigma_{23}$ for $z_{23} = z_{23}^m + \triangle z_{23}$ and $\sigma_{23} = \sigma_{23}^m + \triangle\sigma_{23}$

$$\cos z_{23} = \cos z_{23}^m - \sin z_{23}^m \frac{\triangle z_{23}''}{\varrho''}$$

$$\cot \sigma_{23} = \cot \sigma_{23}^m - \frac{1}{\sin^2 \sigma_{23}^m} \frac{\sigma_{23}''}{\varrho''},$$

it holds that $z_{23} = \cos z_{23}^m + \triangle z_{23}''$ and $\sigma_{23} = cos\sigma_{23}^m + \triangle\sigma_{23}''$.

Equations for reductions of horizontal and zenith angles now can be expressed as

$$z_{23} = \cos z_{23}^m + 2\gamma_{12}'' \cos \sigma_{23}^m$$

$$\sigma_{23} = \sigma_{23}^m - 2\gamma_{12}'' \cot z_{23}^m \sin \sigma_{23}^m.$$

1.4 Reporting bugs

Undoubtedly there are numerous bugs remaining, both in the C++ source code and in the documentation. If you find a bug in either, please send a bug report to

bug-gama@gnu.org

We will try to be as quick as possible in fixing the bugs and redistributing the fixes. If you prefere, you can always write directly to Ale epek.

1.5 Contributors

The following persons (in chronological order) have made contributions to GNU Gama project: Ale epek, Ji Vesel, Petr Doubrava, Jan Pytel, Chuck Ghilani, Dan Haggman, Mauri Visnen, John Dedrum, Jim Sutherland, Zoltan Faludi, Diego Berge, Boris Pihtin, Stphane Kaloustian, Siki Zoltan, Anton Horpynich, Claudio Fontana, Bronislav Koska, Martin Beckett, Ji Novk, Vclav Petr, Jokin Zurutuza, (Vim Xiang) and Tom Kubn.

Vclav Petr is the author of Chapter 3 [SQL schema SQLite and gama-local], page 23.

2 XML input data format for `gama-local`

The input data format for a local geodetic network adjustment (program `gama-local`) is defined in accordance with the definition of Extended Markup Language (XML) for description of structured data. The XML definition can be found at

`http://www.w3.org/TR/REC-xml`

Input data (points, observations and other related information) are described using XML start-end pair tags `<xxx>` and `</xxx>` and empty-element tags `<xxx/>`.

The syntax of XML `gama-local` input format is described in XML schema (XSD), the file `gama-local.xsd` is a part of the `GNU gama` distribution and can formally be validated independently on the program `gama-local`, namely in unit testing we use `xmllint` validating parser, if it is installed.

For parsing the XML input data, `gama-local` uses the XML parser `Expat` copyrighted by James Clark which is described at

`http://www.jclark.com/xml/expat.html`

`Expat` is subject to the Mozilla Public License (MPL), or may alternatively be used under the GNU General Public License (GPL) instead.

In the `gama-local` XML input, distances are given in meters, angular values in centigrades and their standard deviations (rms errors) in millimeters or centigrade seconds, respectively. Alternatively angular values in `gama-local` XML input can be given in degrees and seconds (see Section 2.1 [Angular units], page 7). At the end of this chapter an example of the `gama-local` XML input data object is given.

2.1 Angular units

Horizontal angles, directions and zenith angles in `gama-local` XML adjustment input are implicitly given in gons and their standard deviations and/or variances in centicentigons. Gon, also called centesimal grade and Neugrad (German for new grad), is 1/400-th of the circumference. For example

```
<direction  from="202" to="416" val="63.9347"  stdev="10.0" />
```

The same angular value (direction) can be expressed in degrees as

```
<direction  from="202" to="416" val="57-32-28.428"  stdev="3.24" />
```

In XML adjustment input degrees are coded as a single string, where degrees (57), minutes (32) and seconds (28.428) are separated by dashes (-) with optional leading sign. Spaces are not allowed inside the string. Gons and degrees may be mixed in a single XML document but one should be careful to supply the information on standard deviations and/or covariances in the proper corresponding units.

Internally `gama-local` works with gons but output can be transformed to degrees using the option `--angles 360`.

2.2 Prologue

XML documents begin with an XML declaration that specifies the version of XML being used (*prolog*). In the case of `gama-local` follows the root tag `<gama-local>` with XML Schema namespace defined in attribute `xmlns`:

```
<?xml version="1.0" ?>
<gama-local xmlns="http://www.gnu.org/software/gama/gama-local">
```

GNU Gama uses non-validating parser and the XML Schema Definition namespace is not used in `gama-local` but it is essential for usage in third party software that might need XML validation.

2.3 Tags `<gama-local>` and `<network>`

A pair tag `<gama-local>` contains a single pair tag `<network>` that contains the network definition. The definition of the network is composed of three sections:

- `<description>` of the network (annotation or comments),
- network `<parameters />` and
- `<points-observations>` section.

The sections `<description>` and `<parameters />` are optional, the section `<points-observations>` is mandatory. These three sections may be presented in any order and may be repeated several times (in such a case, the corresponding sections are linked together by the software).

The pair tag `<network>` has two optional attributes `axes-xy` and `angles`. These attributes are used to describe orientation of the `xy` orthogonal coordinate system axes and the orientation of the observed angles and/or directions.

- `axes-xy="ne"` orientation of axes `x` and `y`; value `ne` implies that axis `x` is oriented north and axis `y` is oriented east. Acceptable values are `ne`, `sw`, `es`, `wn` for left-handed coordinate systems and `en`, `nw`, `se`, `ws` for right-handed coordinate systems (default value is `ne`).
- `angles="right-handed"` defines counterclockwise observed angles and/or directions, value `left-handed` defines clockwise observed angles and/or directions (default value is `left-handed`).

Many geodetic systems are right handed with `x` axis oriented east, `y` axis oriented north and counterclockwise angular observations. Example of left-handed orthogonal system with different axes orientation is coordinate system *Krovak* used in the Czech Republic where the axes `x` and `y` are oriented south and west respectively.

GNU Gama can adjust any combination of coordinate and angular systems.

Example

```
<gama-local>
<network>
   <description> ... </description>
   <parameters ... />
   <points-observations> ... </points-observations>
</network>
</gama-local>
```

It is planned in future versions of the program to allow more `<network>` tags (analysis of deformations etc.) and definitions of new tags.

2.4 Network description

The description of a geodetic network is enclosed in the start-end pair tags `<description>`. Text of the description is copied into the adjustment output and serves for easier identification of results. The text is not interpreted by the program, but it may be helpful for users.

Example

```
<description>
A short description of a geodetic network ...
</description>
```

2.5 Network parameters

The network parameters may be listed with the following optional attributes of an empty-element tag `<parameters />`

- `sigma-apr = "10"` value of a priori reference standard deviation—square root of reference variance (default value 10)

- `conf-pr = "0.95"` confidence probability used in statistical tests (dafault value 0.95)

- `tol-abs = "1000"` tolerance for identification of gross absolute terms in project equations (default value 1000 mm)

- `sigma-act = "aposteriori"` actual type of reference standard deviation use in statistical tests (`aposteriori | apriori`); default value is `aposteriori`

- `update-constrained-coordinates = "no"` enables user to control if coordinates of constrained points are updated in iterative adjustment. If test on linerarization fails (see Section 4.9 [Linearization], page 42), Gama tries to improve approximate coordinates of adjusted points and repeats the whole adjustment. Coordinates of constrained points are implicitly not changed during iterations.

- `algorithm = "gso"` numerical algortihm used in the adjistment (gso, svd, cholesky, envelope).

- `languade = "en"` the language to be used in adjustment output.

- `encoding = "utf-8"` adjustment output encoding.

- `angles = "400"` output results angular units (400/360).

- `latitude = "50"`

- `ellipsoid`

- `cov-band = "-1"` the bandwith of covariance matrix of the adjusted parameters in the output XML file (-1 means all covariances).

Values of the attributes must be given either in the double-quotes ("...") or in the single quotes ('...'). There can be *white spaces* (spaces, tabs and new-line characters) between attribute names, values, and the *equal* sign.

Example

```
<parameters sigma-apr = "15"
            conf-pr   = '0.90'
```

```
        sigma-act = "apriori"
        update-constrained-coordinates = "no" />
```

2.6 Points and observations

The points and observations section is bounded by the pair tag `<points-observations>` and contains information about points, observed horizontal directions, angles, and horizontal distances, height differences, slope distances, zenith angles, observed vectors and control coordinates.

Optional attributes of the start tag `<points-observations>` allow for the definition of default values of standard deviations corresponding to observed directions, angles, and distances.

- `direction-stdev = "..."` defines the implicit value of standard deviation of observed directions (default value is not defined)
- `angle-stdev = "..."` defines the implicit value of standard deviation of observed angles (default value is not defined)
- `zenith-angle-stdev = "..."` defines the implicit value of standard deviation of observed zenith angles (default value is not defined)
- `azimuth-stdev = "..."` defines the implicit value of standard deviation of observed azimuth angles (default value is not defined)
- `distance-stdev = "..."` defines the implicit value of standard deviation of observed distances, horizontal or slope (default value is not defined)

Implicit values of standard deviations for the observed distances are calculated from the model with three constants a, b, and c according to the formula

$$a + bD^c,$$

where a is a constant part of the model and D is the observed distance in kilometres. If the constants b and/or c are not given, default values of $b = 0$ and $c = 1$ will be used.

Example

```
<points-observations direction-stdev = "10"
                     distance-stdev  = "5 3 1" >
   <!-- ... points and observation data ... -->
</points-observations>
```

2.7 Points

Points are described by the empty-element tags `<point/>` with the following attributes:

- `id = "..."` is the point identification attribute (mandatory); point identification is not limited to *numbers*; all printable characters can be used in identification.
- `x = "..."` specifies coordinate x
- `y = "..."` specifies coordinate y
- `z = "..."` specifies coordinate z, point height

- `fix = "..."` specifies coordinates that are fixed in adjustment; acceptable values are xy, XY, z, Z, xyz, XYZ, xyZ and XYz.
- `adj = "..."` specifies coordinates to be adjusted (unknown parameters in adjustment); acceptable values are xy, XY, z, Z, xyz, XYZ, xyZ and XYz.

With exception of the first attribute (point id), all other attributes are optional. Decimal numbers can be used as needed.

Control coordinates marked using the `fix` parameter are not changed in the adjustment. Uppercase and lowercase notation of coordinates with the `fix` parameter are interpreted the same. Corrections are applied to the unknown parameters identified by coordinates written in lowercase characters given in the `adj` parameter. When the coordinates are written using uppercase, they are interpreted as *constrained coordinates*. If coordinates are marked with both the `fix` and `adj`, the `fix` parameter will take precedence.

Constrained coordinates are used for the regularization of free networks. If the network is not free (fixed network), the *constrained* coordinates are interpreted as other unknown parameters. In classical free networks, the *constrained* points define the regularization constraint

$$\sum dx_i^2 + dy_i^2 = \min.$$

where dx and dy are adjusted coordinate corrections and the summation index i goes over all *constrained* points. In other words, the set of the *constrained* points defines the adjustment of the free network (its shape and size) with a simultaneous transformation to the approximate coordinates of selected points. Program `gama-local` allows the definition of constrained coordinates with 1D leveling networks, 2D and 3D local networks.

Example

```
<point id="1" y="644498.590" x="1054980.484" fix="xy"  />
<point id="2" y="643654.101" x="1054933.801" adj="XY" />
<point id="403" adj="xy" />
```

2.8 Set of observations

The pair tag `<obs>` groups together a set of observations which are somehow related. A typical example is a set of directions and distances observed from one stand-point. An observation section contains a set of

- horizontal directions `<direction ... />`
- horizontal distances `<distance ... />`
- horizontal angles `<angle ... />`
- slope distances `<s-distance ... />`
- zenith angles `<z-angle ... />`
- azimuths `<azimuth ... />`

The band variance-covariance matrix of directions, distances, angles or other observations listed in one `<obs>` section may be supplied using a `<cov-mat>` pair tag with attributes `dim` (dimension) and `band` (bandwidth). The band-width of the diagonal matrix is equal to 0 and a fully-populated variance-covariance matrix has a bandwidth of `dim-1`.

Observation variances and covariances (i.e. an upper-symmetric part of the band-matrix) are written row by row between `<cov-mat>` and `</cov-mat>` tags. If present, the dimension of the variance-covariance matrix must agree with the number of observations.

The following example of variance-covariance matrix with dimension 6 and bandwidth 2 (two nonzero codiagonals and three zero codiagonals)

$$\begin{pmatrix} 1.1 & 0.1 & 0.2 & 0 & 0 & 0 \\ 0.1 & 1.2 & 0.3 & 0.4 & 0 & 0 \\ 0.2 & 0.3 & 1.3 & 0.5 & 0.6 & 0 \\ 0 & 0.4 & 0.5 & 1.4 & 0.7 & 0.8 \\ 0 & 0 & 0.6 & 0.7 & 1.5 & 0.9 \\ 0 & 0 & 0 & 0.8 & 0.9 & 1.6 \end{pmatrix}$$

is coded in XML as

```
<cov-mat dim="6" band="2">
    1.1  0.1  0.2
         1.2  0.3  0.4
              1.3  0.5  0.6
                   1.4  0.7  0.8
                        1.5  0.9
                             1.6
</cov-mat>
```

If two or more sets of directions with different orientations are observed from a stand-point, they must be placed in different `<obs>` sections. The value of an orientation angle can be explicitly stated with an attribute `orientation="..."`. Normally, it is more convenient to let the program calculate approximate values of orientations needed for the adjustment. If directions are present, then the attribute `station` must be defined.

Optional attribute `from_dh="..."` enables to enter implicit height of instrument for all observations within the `<obs>` pair tag.

Observed distances are expressed in meters, their standard deviations in millimeters. Observed directions and angles are expressed in centigrades (400) and their standard deviations in centigrade seconds.

Height differences can be entered in the `<obs>` or `<height-differences>` section. If entered in the `<obs>` section, the `dist="..."` parameter is ignored (Section 2.15 [Height differences], page 16).

Example

```
<obs from="418">
   <direction  to=  "2" val="0.0000"    stdev="10.0" />
   <direction  to="416" val="63.9347"   stdev="10.0" />
   <direction  to="420" val="336.3190"  stdev="10.0" />
   <distance   to="420" val="246.594"   stdev="5.0"  />
</obs>

<obs from="418">
```

```
<direction  to=  "2" val="0.0000"   />
<direction  to="416" val="63.9347"  />
<direction  to="420" val="336.3190" />
<distance   to="420" val="246.594"  />

<cov-mat dim="4" band="0">
    100.00 100.00 100.00 25.00
</cov-mat>
</obs>
```

2.9 Directions

Directions are expressed with the following attributes in an empty-element tag `<direction />`

- `to = "..."` target point identification
- `val = "..."` observed direction; see Section 2.1 [Angular units], page 7
- `stdev = "..."` standard deviation (optional)
- `from_dh = "..."` instrument height (optional)
- `to_dh = "..."` reflector/target height (optional)

The standard deviation is an optional attribute. However since all observations in the adjustment must have their weights defined, the standard deviation must be given either explicitly with the attribute `stdev="..."` or implicitly with `<points-observation direction-stdev="..." >` or with a variance-covariance matrix for the given observation set. A similar approach applies to all the observations (distances, angles, etc.)

Example

```
<direction  to=  "2" val="0.0000"  stdev="10.0" />
<direction  to="416" val="63.9347" />
```

2.10 Horizontal distances

Distances are written using an empty-element tag `<distance />` with attributes

- `from = "..."` standpoint identification
- `to = "..."` target identification
- `val = "..."` observed horizontal distance
- `stdev = "..."` standard deviation of observed horizontal distance (optional)
- `from_dh = "..."` instrument height (optional)
- `to_dh = "..."` reflector/target height (optional)

Contrary to directions, distances in an observation set (`<obs>`) do not need to share a common stand-point. An example is set of distances observed from several stand-points with a common variance-covariance matrix.

Example

```
<distance from = "2"  to = "1" val = "659.184" />
<distance to ="422" val="228.207"  stdev="5.0"  />
<distance to ="408" val="568.341" />
```

2.11 Angles

Observed angles are expressed with the following attributes of an empty-element tag `<angle />`

- `from = "..."` standpoint identification (optional)
- `bs = "..."` backsight target identification
- `fs = "..."` foresight target identification
- `val = "..."` observed angle; see Section 2.1 [Angular units], page 7
- `stdev = "..."` standard deviation (optional)
- `from_dh = "..."` instrument height (optional)
- `bs_dh = "..."` backsight reflector/target height (optional)
- `fs_dh = "..."` foresight reflector/target height (optional)

Similar to distance observations, one observation set may group angles observed from several standpoints.

Example

```
<angle from="433" bs="422" fs="402" val="128.6548" stdev="14.1"/>
<angle from="433" bs="422" fs="402" val="128.6548" />
<angle bs="422" fs="402" val="128.6548" stdev="14.1"/>
<angle bs="422" fs="402" val="128.6548"/>
```

2.12 Slope distances

Slope distances (space distances) are written using an empty-element tag `<s-distance />` with attributes

- `from = "..."` standpoint identification (optional)
- `to = "..."` target identification
- `val = "..."` observed slope distance
- `stdev = "..."` standard deviation of observed slope distance (optional)
- `from_dh = "..."` instrument height (optional)
- `to_dh = "..."` reflector/target height (optional)

Similar to horizontal distances, one observation set may group slope distances observed from several standpoints.

Example

```
<s-distance from = "2"  to = "1" val = "658.824" />
<s-distance to ="422" val="648.618"  stdev="5.0"  />
<s-distance to ="408" val="482.578" />
```

2.13 Zenith angles

Zenith angles are written using an empty-element tag `<z-angle />` with the following attributes

- `from = "..."` standpoint identification (optional)
- `to = "..."` target identification
- `val = "..."` observed zenith angle; see Section 2.1 [Angular units], page 7
- `stdev = "..."` standard deviation of observed zenith angle (optional)
- `from_dh = "..."` instrument height (optional)
- `to_dh = "..."` reflector/target height (optional)

Similar to horizontal distances, one observation set may group zenith angles observed from several standpoints.

Example

```
<z-angle from = "2"  to = "1" val = "79.6548" />
<z-angle to ="422" val="85.4890"  stdev="5.0"  />
<z-angle to ="408" val="95.7319" />
```

2.14 Azimuths

The azimuth is defined in GNU Gama as an observed horizontal angle measured from the North to the given target. The true north orientation is measured by gyrotheodolites, mainly in mine surveying. In Gama azimuths' angle can be measured clockwise or counterclocwise according to the angle orientation defined in `<parameters />` tag.

Azimuths are expressed with the following attributes in an empty-element tag `<azimuth />`

- `from = "..."` standpoint identification
- `to = "..."` target point identification
- `val = "..."` observed azimuth; see Section 2.1 [Angular units], page 7
- `stdev = "..."` standard deviation (optional)
- `from_dh = "..."` instrument height (optional)
- `to_dh = "..."` reflector/target height (optional)

The standard deviation is an optional attribute. However since all observations in the adjustment must have their weights defined, the standard deviation must be given either explicitly with the attribute `stdev="..."` or implicitly with `<points-observation azimuth-stdev="..." >` or with a variance-covariance matrix for the given observation set.

Example

```
<points-observations azimuth-stdev="15.0">

<azimuth from="1"  to=  "2" val= "96.484371" />
```

2.15 Height differences

A set of observed leveling height differences is described using the start-end tag
`<height-differences>` without parameters. The `<height-differences>` tag can
contain a series of height differences (at least one) and can optionally be supplied with a
variance-covariance matrix. Single height differences are defined with empty tags `<dh />`
having the following attributes:

- `from = "..."` standpoint identification
- `to = "..."` target identification
- `val = "..."` observed leveling height difference
- `stdev = "..."` standard deviation of levellin elevation and
- `dist = "..."` distance of leveling section (in kilometers)

If the value of standard deviation is not present and length of leveling section (in kilometres)
is defined, the value of standard deviation is computed from the formula

$$m_{dh} = m_0 \sqrt{D_{km}}.$$

If the value of standard deviation of the height difference is defined, information on leveling
section length is ignored. A third possibility is to define a common variance-covariance
matrix for all elevations in the set.

Example

```
<height-differences>
  <dh from="A" to="B" val=" 25.42" dist="18.1" />
  <dh from="B" to="C" val=" 10.34" dist=" 9.4" />
  <dh from="C" to="A" val="-35.20" dist="14.2" />
  <dh from="B" to="D" val="-15.54" dist="17.6" />
  <dh from="D" to="E" val=" 21.32" dist="13.5" />
  <dh from="E" to="C" val="  4.82" dist=" 9.9" />
  <dh from="E" to="A" val="-31.02" dist="13.8" />
  <dh from="C" to="D" val="-26.11" dist="14.0" />
</height-differences>
```

2.16 Control coordinates

Control (known) coordinates are described by the start-end pair tag `<coordinates>`. A series of points with known coordinates can be defined using the `<point />` tag. The variance-covariance matrix for the entire set of points can be created with a single `<cov-mat>` tag. In the `<point />` tags, a point identification (ID) and its coordinates (x, y and z) must be listed. Although the order of the `<point />` tag attributes is irrelevant in the corresponding variance-covariance matrix, the expected order of the coordinates is x, y and z (the horizontal coordinates x, y, or the height z might be missing, but not both). The type of the points may be defined either directly within the `<coordinates>` tag or outside of it.

Example

```
<coordinates>
    <point id="1" x="100.00" y="100.00" />
    <point id="2" z="200.00" y="200.00" x="200.00" />
    <point id="3" z="300.00" />
    <cov-mat dim="6" band="5" >
        ...  <!-- covariances for 1x 1y 2x 2y 2z 3z -->
    </cov-mat>
</coordinates>
```

2.17 Coordinate differences (vectors)

Observed coordinate differences describe relative positions of station pairs (vectors). Contrary to the observed coordinates, the variance-covariance matrix of the coordinate differences always describes all three elements of the 3D vectors.

Optional attributes of empty element tag `<vec>` for describing instrument and/or target height are

- `from_dh = "..."` instrument height
- `to_dh = "..."` target height

Example

```
<vectors>
    <vec from="id1" to="id2" dx="..." dy="..." dz="..." />
    <vec from="id2" to="id3" dx="..." dy="..." dz="..." />
    ...
    <cov-mat dim="..." band="..." >
        ..
    </cov-mat>
</vectors>
```

2.18 Example of local geodetic network

The XML input data format should be now reasonably clear from the following sample geodetic network. This example is taken from user's guide to Geodet/PC by Frantisek Charamza.

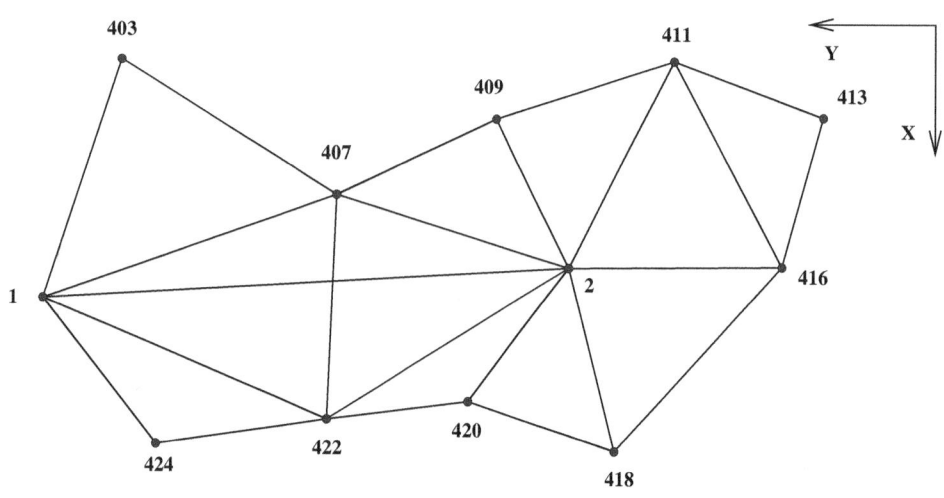

```
<?xml version="1.0" ?>

<gama-local xmlns="http://www.gnu.org/software/gama/gama-local">
<network axes-xy="sw">

<description>
XML input stream of points and observation data for the program GNU gama
</description>

<!-- parameters are expressed with empty-element tag -->

<parameters sigma-act = "aposteriori" />

<points-observations>

<!-- fixed point, constrained point -->

<point id="1" y="644498.590" x="1054980.484" fix="xy" />
<point id="2" y="643654.101" x="1054933.801" adj="XY" />

<!-- computed / adjusted points -->

<point id="403" adj="xy" />
<point id="407" adj="xy" />
<point id="409" adj="xy" />
```

```
<point id="411" adj="xy" />
<point id="413" adj="xy" />
<point id="416" adj="xy" />
<point id="418" adj="xy" />
<point id="420" adj="xy" />
<point id="422" adj="xy" />
<point id="424" adj="xy" />

<obs from="1">
     <direction  to=  "2" val=  "0.0000" stdev="10.0" />
     <direction  to="422" val= "28.2057" stdev="10.0" />
     <direction  to="424" val= "60.4906" stdev="10.0" />
     <direction  to="403" val="324.3662" stdev="10.0" />
     <direction  to="407" val="382.8182" stdev="10.0" />
     <distance   to=  "2" val= "845.777" stdev="5.0"  />
     <distance   to="422" val= "493.793" stdev="5.0"  />
     <distance   to="424" val= "288.301" stdev="5.0"  />
     <distance   to="403" val= "388.536" stdev="5.0"  />
     <distance   to="407" val= "498.750" stdev="5.0"  />
</obs>

<obs from="2">
     <direction  to=  "1" val="0.0000"   stdev="10.0" />
     <direction  to="407" val="22.2376"  stdev="10.0" />
     <direction  to="409" val="73.8984"  stdev="10.0" />
     <direction  to="411" val="134.2090" stdev="10.0" />
     <direction  to="416" val="203.0706" stdev="10.0" />
     <direction  to="418" val="287.2951" stdev="10.0" />
     <direction  to="420" val="345.6928" stdev="10.0" />
     <direction  to="422" val="368.9908" stdev="10.0" />
     <distance   to="407" val="388.562"  stdev="5.0"  />
     <distance   to="409" val="257.498"  stdev="5.0"  />
     <distance   to="411" val="360.282"  stdev="5.0"  />
     <distance   to="416" val="338.919"  stdev="5.0"  />
     <distance   to="418" val="292.094"  stdev="5.0"  />
     <distance   to="420" val="261.408"  stdev="5.0"  />
     <distance   to="422" val="452.249"  stdev="5.0"  />
</obs>

<obs from="403">
     <direction  to=  "1" val="0.0000"   stdev="10.0" />
     <direction  to="407" val="313.5542" stdev="10.0" />
     <distance   to="407" val="405.403"  stdev="5.0"  />
</obs>

<obs from="407">
     <direction  to=  "1" val="0.0000"   stdev="10.0" />
```

```
        <direction   to="403" val="55.1013"   stdev="10.0" />
        <direction   to="409" val="193.3410"  stdev="10.0" />
        <direction   to=  "2" val="239.4204"  stdev="10.0" />
        <direction   to="422" val="323.5443"  stdev="10.0" />
        <distance    to="409" val="281.997"   stdev="5.0"  />
        <distance    to="422" val="346.415"   stdev="5.0"  />
</obs>

<obs from="409">
        <direction   to=  "2" val="0.0000"    stdev="10.0" />
        <direction   to="407" val="102.2575"  stdev="10.0" />
        <direction   to="411" val="310.1751"  stdev="10.0" />
        <distance    to="411" val="296.281"   stdev="5.0" />
</obs>

<obs from="411">
        <direction   to=  "2" val="0.0000"    stdev="10.0" />
        <direction   to="409" val="49.8647"   stdev="10.0" />
        <direction   to="413" val="291.4953"  stdev="10.0" />
        <direction   to="416" val="337.6667"  stdev="10.0" />
        <distance    to="413" val="252.266"   stdev="5.0"  />
        <distance    to="416" val="360.449"   stdev="5.0"  />
</obs>

<obs from="413">
        <direction   to="411" val="0.0000"    stdev="10.0" />
        <direction   to="416" val="295.3582"  stdev="10.0" />
        <distance    to="416" val="239.745"   stdev="5.0"  />
</obs>

<obs from="416">
        <direction   to=  "2" val="0.0000"    stdev="10.0" />
        <direction   to="411" val="68.8065"   stdev="10.0" />
        <direction   to="413" val="117.9922"  stdev="10.0" />
        <direction   to="418" val="348.1606"  stdev="10.0" />
        <distance    to="418" val="389.397"   stdev="5.0"  />
</obs>

<obs from="418">
        <direction   to=  "2" val="0.0000"    stdev="10.0" />
        <direction   to="416" val="63.9347"   stdev="10.0" />
        <direction   to="420" val="336.3190"  stdev="10.0" />
        <distance    to="420" val="246.594"   stdev="5.0"  />
</obs>

<obs from="420">
        <direction   to=  "2" val="0.0000"    stdev="10.0" />
```

```
        <direction  to="418" val="77.9221"  stdev="10.0" />
        <direction  to="422" val="250.1804" stdev="10.0" />
        <distance   to="422" val="228.207"  stdev="5.0"  />
</obs>

<obs from="422">
        <direction  to=  "2" val="0.0000"   stdev="10.0" />
        <direction  to="420" val="26.8834"  stdev="10.0" />
        <direction  to="424" val="225.7964" stdev="10.0" />
        <direction  to=  "1" val="259.2124" stdev="10.0" />
        <direction  to="407" val="337.3724" stdev="10.0" />
        <distance   to="424" val="279.405"  stdev="5.0"  />
</obs>

<obs from="424">
        <direction  to=  "1" val="0.0000"   stdev="10.0" />
        <direction  to="422" val="134.2955" stdev="10.0" />
</obs>

</points-observations>

</network>
</gama-local>
```

3 SQL schema, SQLite and `gama-local`

The input data for a local geodetic network adjustment (program `gama-local`) can be stored in SQLite 3 database file. The general information about SQLite can be found at

http://www.sqlite.org/

Input data (points, observations and other related information) are stored in SQLite database file. Native SQLite C/C++ API is used for reading SQLite database file. It is described at

http://www.sqlite.org/c3ref/intro.html

Please note if you compile GNU Gama as described in Section 1.2 [Install], page 2 and SQLite library is not installed on your system, GNU Gama would be compiled without SQLite support.

SQL schema (`CREATE` statements) is in `gama-local-schema.sql` file which is part of GNU Gama distribution and is in the `xml` directory.

All tables for `gama-local` are prefixed with `gnu_gama_local_`. In the documentation table names are referred without this prefix. For example table `gnu_gama_local_points` is referred as `points`.

Database scheme used for SQLite database is also valid in other SQL database systems. Almost every column has some constraint to ensure correctness.

You can convert existing XML input file to SQL commands with program `gama-local-xml2sql`, for example

```
$ gama-local-xml2sql geodet-pc geodet-pc-123.gkf geodet-pc.sql
```

3.1 Working with SQLite database

First of all you have to create tables for GNU Gama in SQLite database file (here with `db` extension, but you can choose your own, e.g. `sqlite`).

```
$ sqlite3 gama.db < gama-local-schema.sql
```

You can check created tables by following commands (fist in command line, second in SQLite command line).

```
$ sqlite3 gama.db
sqlite> .tables
```

Output should look like this:

```
gnu_gama_local_clusters        gnu_gama_local_descriptions
gnu_gama_local_configurations  gnu_gama_local_obs
gnu_gama_local_coordinates     gnu_gama_local_points
gnu_gama_local_covmat          gnu_gama_local_vectors
```

When you have created tables you can import data. One way is to process file with SQL statements.

```
$ sqlite3 gama.db < geodet-pc.sql
```

Another way can be filing database file in another program.

For using `sqlite3` command you need a command line interface for SQLite 3 installed on your system (e.g. `sqlite3` package).

3.2 Units in SQL tables

In the `gama-local` SQLite database, distances are given in meters and their standard deviations (rms errors) in millimeters. Angular values are given in radians as well as their standard deviations.

Conversions between radians, gons and degrees:

$$\mathrm{rad} = \mathrm{gon} \cdot \frac{\pi}{200} = \mathrm{deg} \cdot \frac{\pi}{180}$$

3.3 Network SQL definition

Network definitions are stored in the `configurations` table. This table contains all parameters for each network such as value of a priori reference standard deviation or orientation of the `xy` orthogonal coordinate system axes.

It is obvious that in one database file can be stored more networks (configurations).

Configuration descriptions (annotation or comments) are stored separately in table `descriptions`. The description is split to many records because of compatibility with various databases (not all databases implements type `TEXT`).

Field (attribute) `conf_id` identifies a configuration in the database. Field `conf_name` is used to identify configuration outside the database (e.g. parameter in command-line when reading data from database to `gama-local`).

Table `configurations` contains all parameters specified in tag `<parameters />` (see Section 2.5 [Network parameters], page 9) and also `gama-local` command line parameters (see Section 1.3 [Program gama-local], page 3). The list of all table attributes (parameters) follows.

- `sigma_apr` value of a priori reference standard deviation—square root of reference variance (default value 10)

- `conf_pr` confidence probability used in statistical tests (dafault value 0.95)

- `tol_abs` tolerance for identification of gross absolute terms in project equations (default value 1000 mm)

- `sigma_act` actual type of reference standard deviation use in statistical tests (`aposteriori | apriori`); default value is `aposteriori`

- `update_cc` enables user to control if coordinates of constrained points are updated in iterative adjustment. If test on linerarization fails (see Section 4.9 [Linearization], page 42), Gama tries to improve approximate coordinates of adjusted points and repeats the whole adjustment. Coordinates of constrained points are implicitly not changed during iterations. Acceptable values are `yes`, `no`, default value is `no`.

- `axes_xy` orientation of axes `x` and `y`; value `ne` implies that axis `x` is oriented north and axis `y` is oriented east. Acceptable values are `ne`, `sw`, `es`, `wn` for left-handed coordinate systems and `en`, `nw`, `se`, `ws` for right-handed coordinate systems (default value is `ne`).

- `angles right-handed` defines counterclockwise observed angles and/or directions, value `left-handed` defines clockwise observed angles and/or directions (default value is `left-handed`).

- `epoch` is measurement epoch. It is floating point number (default value is `0.0`).

- `algorithm` specifies numerical method used for solution of the adjustment. For Singular Value Decomposition set value to `svd`. Value `gso` stands for block matrix algorithm GSO by Frantisek Charamza based on Gram-Schmidt orthogonalization, value `cholesky` for Cholesky decomposition of semidefinite matrix of normal equations and value `envelope` for a Cholesky decomposition with *envelope* reduction of the sparse matrix. Default value is `svd`.

- `ang_units` Angular units of angles in `gama-local` output. Value 400 stands for gons and value 360 for degrees (default value is 400). Note that this doesn't effect units of angles in database. For further information about angular units see Section 2.1 [Angular units], page 7.

- `latitude` is mean latitude in network area. Default value is 50 (gons).

- `ellipsoid` is name of ellipsoid (see Section 5.2 [Supported ellipsoids], page 47).

All fields are mandatory except `ellipsoid` field. For additional information about handling geodetic systems in `gama-local` see Section 2.3 [Network definition], page 8.

Example (`configuration` table contents):

```
conf_id|conf_name|sigma_apr|conf_pr|tol_abs|sigma_act  |update_cc|...
----------------------------------------------------------------------
1      |geodet-pc|10.0     |0.95   |1000.0 |aposteriori|no       |...


... axes_xy|angles        |epoch|algorithm|ang_units|latitude|ellipsoid
----------------------------------------------------------------------
... ne     |left-handed|0.0 |svd      |400      |50.0    |
```

The list of `description` table attributes follows.

- `conf_id` is id of configuration which description (text) belongs to.

- `id` identifies text in a database.

- `text` is part of configuration description. Its SQL type is `VARCHAR(1000)`.

There can be more than one text for one configuration. All texts related to one configuration are concatenated to one description.

Example (`description` table contents):

```
conf_id|indx|text
------------------------------------------------
1      |1   |Frantisek Charamza: GEODET/PC, ...
```

3.4 Table `points`

- `conf_id` is id of configuration which points belongs to.

- `id` identifies point in a database and also in an output. It is mandatory and it is character string (SQL type is `VARCHAR(80)`). Point `id` has to be unique within one configuration. In documentation it is referred as point identification or point id.

- `x`, `y` and `z` coordinates of a point. Coordinate `z` is considered as height.

- `txy` and `tz` specify the type of coordinates `x`, `y` and `z`. Acceptable values are `fixed`, `adjusted` and `constrained` (there is no default value). For details see Section 2.7 [Points], page 10.

Example (table contents):

```
conf_id|id |x         |y       |z|txy     |tz

----------------------------------------
1       |201|78594.91|9498.26| |fixed   |
1       |205|78907.88|7206.65| |fixed   |
1       |206|76701.57|6633.27| |fixed   |
1       |207|        |        | |adjusted|
```

3.5 Table clusters

The cluster is a group of observations with the common covariance matrix. The covariance matrix allows to express any combination of correlations among observations in cluster (including uncorrelated observations, where covariance matrix is diagonal). For explanation see Section 5.1 [Observation data and points], page 45.

In the database observations are stored in three tables: obs, coordinates and vectors. Cluster's covariance matrix is stored in table covmat. Every observation, vector or coordinate in database has to be in some cluster.

- conf_id is id of configuration which cluster belongs to.

- ccluster identifies a cluster within one configuration.

- dim and band specify dimension and bandwidth of covariance matrix. The bandwidth of the diagonal matrix is equal to 0 and a fully-populated covariance matrix has a bandwidth of dim-1 (band maximum possible value is dim-1).

- tag specifies type of observations in cluster which also implies the table where they are stored in. obs and height-differences stand for obs table, coordinates and vectors stand for coordinates table and vectors table respectively.

Observations, vectors and coordinates are identified by configuration id (conf_id), cluster id ccluster and theirs index (indx). Observation index (indx) has to be unique within observations of one cluster (which belongs to one configuration). The same applies for vectors and coordinates.

See also Section 2.8 [Set of observations], page 11.

Example (table contents):

```
conf_id|ccluster|dim|band|tag

------------------------------
1       |1       |3  |0   |obs
1       |4       |4  |0   |obs
```

3.6 Table covmat

Values of cluster covariance matrix are stored in covmat table. Attributes conf_id, ccluster identifies covariance matrix. Value position in matrix is specified by rind and cind fields.

- conf_id is id of configuration which cluster belongs to.

- ccluster is id of cluster which matrix belongs to.

- rind is row number in covariance matrix

- `cind` is column number covariance matrix
- `val` is value itself (variance or covariance).

Values `rind` and `cind` have to respect `dim` and `band` specified in table `clusters`. If value in covariance matrix is not specified (record is missing), it is considered to be zero.

Example (table contents):

```
conf_id|ccluster|rind|cind|val
-------------------------------
1       |1       |1   |1   |400.0
1       |1       |2   |2   |400.0
1       |1       |3   |3   |400.0
1       |4       |1   |1   |400.0
1       |4       |2   |2   |400.0
1       |4       |3   |3   |400.0
1       |4       |4   |4   |400.0
```

3.7 Table `obs`

Table `obs` contains simple observations like direction or distance.

- `conf_id` is id of configuration which cluster belongs to.
- `ccluster` is id of cluster which observation belongs to.
- `indx` identifies observation within cluster. It has to be positive integer.
- `tag` specifies a type of an observation. Allowed `tags` follows.
  - `direction` for directions.
  - `distance` for horizontal distances.
  - `angle` for angles.
  - `s-distance` for slope distances (space distances).
  - `z-angle` for zenith angles.
  - `azimuth` for azimuth angles.
  - `dh` for leveling height differences.
- `from_id` is stand point identification. It is mandatory and it must not differ within one cluster for observations with `tag = 'direction'` .
- `to_id` is target identification (mandatory).
- `to_id2` is second target identification. It is valid and mandatory only for angles (`tag = 'angle'`).
- `val` is observation value. It is mandatory for all observation types.
- `stdev` is value of standard deviation. It is used when variance in covariance matrix is not specified.
- `from_dh` is value of instrument height (optional).
- `to_dh` is value of reflector/target height (optional).
- `to_dh2` is value of second reflector/target height (optional). It is valid only for angles.
- `dist` is distance of leveling section. It is valid only for height-differences (`tag = 'dh'`).

- **rejected** specifies whether observation is rejected (passive) or not. Value 0 stand for not rejected, value 1 for rejected. It is mandatory. Default value is 0.

Example (table contents without empty columns):

```
conf_id|ccluster|indx|tag       |from_id|to_id|val             |rejected
-----------------------------------------------------------------------
   1    |1       |1   |direction|201    |202  |0.0             |0
   1    |1       |2   |direction|201    |207  |0.817750284544|0
   1    |1       |3   |direction|201    |205  |2.020073921388|0
```

3.8 Table coordinates

Table `coordinates` contains control (known) coordinates.

- **conf_id** is id of configuration which cluster belongs to.
- **ccluster** is id of cluster which coordinates belongs to.
- **indx** identifies coordinates within cluster. It has to be positive integer.
- **id** is point identification.
- **x**, **y** and **z** are coordinates.
- **rejected** specifies whether observation is rejected (passive) or not. Value 0 stand for not rejected, value 1 for rejected. Default value is 0.

See also Section 2.16 [Control coordinates], page 17.

3.9 Table vectors

Table `vectors` contains coordinate differences (vectors).

- **conf_id** is id of configuration which cluster belongs to.
- **ccluster** is id of cluster which vector belongs to.
- **indx** identifies vector within cluster. It has to be positive integer.
- **from_id** is point identification. It identifies initial point.
- **to_id** is point identification. It identifies terminal point.
- **dx**, **dy** and **dz** are coordinate differences.
- **from_dh** is value of initial point height. It is optional.
- **to_dh** is value of terminal point height. It is optional.
- **rejected** integer default 0 not null,

See also Section 2.17 [Coordinate differences], page 17.

3.10 Example of local geodetic network in SQL

Providing complete example would be reasonable because of its extent. However, you can obtain example by following these instructions:

Create a file with XML representation of network by copy and paste example from Section 2.18 [Example], page 18 to a new file. Note that file should start with `<?xml version="1.0" ?>` (no whitespace). Alternatively you can use existing XML file from collection of sample networks (see Section 1.1 [Download], page 2). Then you can convert your XML file (here `example_network.xml`) to SQL statements by program `gama-local-xml2sql` (the path depends on your Gama installation).

```
$ gama-local-xml2sql example_net example_network.xml example_network.sql
```

Now you have example network (configuration `example_net`) in the form of SQL `INSERT` statements in the file `example_network.sql`.

Another representations you can create and fill SQLite database (for details see Section 3.1 [Working with SQLite database], page 23):

```
$ sqlite3 examples.db < gama-local-schema.sql
$ sqlite3 examples.db < example_network.sql
$ sqlite3 examples.db
```

Once you have SQLite database, you can work with it from SQLite command line. You can get nice output by executing following commands.

```
sqlite> .mode column
sqlite> .nullvalue NULL
sqlite> SELECT * FROM gnu_gama_local_configurations;
sqlite> SELECT * FROM gnu_gama_local_points;
sqlite> SELECT * FROM gnu_gama_local_clusters;
sqlite> SELECT * FROM gnu_gama_local_covmat;
sqlite> SELECT * FROM gnu_gama_local_obs;
```

Or you can get database dump (`CREATE` and `INSERT` statements) by

```
sqlite> .dump
```

If it is not enough for you, you can try one of GUI tools for SQLite.

4 Network adjustment with `gama-local`

Adjustment of local geodetic network is a classical case of *adjustment of indirect observations*. After estimation of approximate values of unknown parameters (coordinates of points) and linearization of functions describing relations between observations and parameters we solve linear system of equations

$$\mathbf{Ax} = \mathbf{b} + \mathbf{v}, \tag{1}$$

where $\mathbf{A}$ is coefficient matrix, $\mathbf{b}$ is vector of absolute terms (right hand side) and $\mathbf{v}$ is vector of residuals. This system is (generally) overdetermined and we seek the solution $\mathbf{x}$ satisfying the basic criterion of Least Squares

$$\mathbf{v'Pv} = \min, \tag{2}$$

where $\mathbf{P}$ is weight matrix. This criterion unambiguously defines the shape of adjusted network.

In the case of *free network* the system (1) is singular (matrix $\mathbf{A}$ has linearly dependent columns) and we have to define second regularization criterion

$$\sum_{i \in \Omega} x_i^2 = \min, \tag{3}$$

stating that at the same time we demand that the sum of squares corrections of selected parameters is minimal (corrections of unknown parameters with indexes from the set Ω). Geometrically this criterion is equivalent to adjustment of the network according to (2) with simultaneous transformation to the selected set of fiducial points. This transformation does not change the shape of adjusted network.

Often it is advantageous to work with a *homogenized system,* ie. with the system of project equations in which coefficient of each row and absolute term are multiplied by square root of the weight of corresponding observation.

$$\mathbf{\tilde{A}x} = \mathbf{\tilde{b}}, \tag{4}$$

where $\mathbf{\tilde{A}} = \mathbf{P}^{1/2}\mathbf{A}$, $\mathbf{\tilde{b}} = \mathbf{P}^{1/2}\mathbf{b}$. Symbol $\mathbf{P}^{1/2}$ denotes diagonal matrix of square roots of observation weights (or Cholesky decomposition of covariance matrix in the case of correlated observations). To criterion (2) corresponds in the case of homogenized system criterion

$$\mathbf{\tilde{v}'v} = \min. \tag{5}$$

Normal equations are clearly equivalent for both systems.

$$(\mathbf{A'PA})\mathbf{x} = (\mathbf{A'Pb}) \quad \equiv \quad (\mathbf{\tilde{A}'\tilde{A}})\mathbf{x} = (\mathbf{\tilde{A}\tilde{b}}).$$

Between weight coefficients of the original system (1) and homogenized system (4) are the following relations

$$q_{x_i} = \tilde{q}_{x_i}, \qquad i = 1, \ldots, n,$$
$$q_{L_j} = \tilde{q}_{L_j}/p_j, \qquad j = 1, \ldots, m,$$
$$q_{v_k} = \tilde{q}_{v_k}/p_k = (1 - \tilde{q}_{L_k})/p_k = 1/p_k - q_{L_k}, \qquad k = 1, \ldots, m.$$

4.1 Approximate coordinates

For computation of coefficients in system (1) (ie. during linearization) we need, first of all, an estimate of approximate coordinates of points and approximate values of orientations of observed directions sets.

Approximate values of unknown parameters are usually not known and we have to compute them from the available observations. For approximate value of orientation program `gama-local` uses median of all estimates from the given set of directions to the points with known coordinates. Median is less sensitive to outliers than arithmetic mean which is normally used for approximate estimate of orientations

During the phase of computation of approximate coordinate of points, program `gama-local` walks through the list of computed points and for each point gathers all determining elements pointing to points with known or previously computed coordinates. Determining elements are

> **outer bearing** (oriented half-line) starting from the point with known coordinates and pointing to the computed point
>
> **distance** between given and computed points
>
> **inner angle** with vertex in the computed point and arms intersecting given points

For all combinations of determining elements program `gama-local` computes intersections and estimates approximate coordinates as the median of all available solutions.

If at least one point was resolved while iterating through the list, the whole cycle is repeated.

If no more coordinates can be solved using intersections and points with unknown coordinates are remaining, program tries to compute coordinates of unresolved points in a local coordinates system and obtain their coordinates using similarity transformation. If a transformation succeeds to resolve coordinates at least one computed point and there are still some points without coordinates left, the whole process is repeated. Classes for computation of approximate coordinates have been written by Jiri Vesely.

If program `gama-local` fails to compute approximate coordinates of some of the network points, they are eliminated from the adjustment and they are listed in the output listing.

With the outlined strategy, program `gama-local` is able to estimate approximate coordinates in most of the cases we normally meet in surveying profession. Still there are cases in which the solution fails. One example is an inserted horizontal traverse with sets of observed direction on both ends but without a connecting observed distance. The solution of approximate coordinates can fail when there is a number of gross error for example resulting from confusion of point identifications but in normal situations, leaving computation of approximate coordinates on program `gama-local` is recommended.

Example

```
Computation of approximate coordinates of points
*************************************************

Number of points with given coordinates:     2
Number of solved points             :        2
Number of observations              :        4
```

```
----------------------------------------------------------
Successfully solved points          :      0
Remaining unsolved points           :      2


List of unresolved points
**************************
422
424
```

4.2 Gross absolute terms

One of parameters in XML input of program `gama-local` is tolerance `tol-abs` for detecting of gross absolute terms in project equations. Observations with outlying absolute terms are always excluded from adjustment.

For measured distances program tests difference between observed value d_i and distance computed from approximate coordinates d_0

$$|d_i - d_0| > \texttt{tol} - \texttt{abs},$$

for observed directions program `gama-local` tests transverse deviation corresponding to absolute term b_i from project equations (1)

$$|b_i| d_0 > \texttt{tol} - \texttt{abs}$$

and similarly for angles, program tests the greater of two deviations corresponding to left and right distances (left and right arm of the angle)

$$|b_i| \max\{d_{0_l}, d_{0_r}\} > \texttt{tol} - \texttt{abs}.$$

Default value of parameter `tol-abs` is 1000 mm.

Example

```
Outlying absolute terms in project equations
*********************************************

   i    standpoint        target           observed      absolute
============================================== value ===== term ==

   2          103         104 dir.      301.087900        -9989.1


Observations with outlying absolute terms removed
```

4.3 Parameters of statistical analysis

Program `gama-local` uses two basic statistical parameters

- confidence probability P (default value is 95%, see input XML parameter `conf-pr`) and

- actual type of reference standard deviation m_{0a} (parameter `sigma-act`).

Confidence probability determines significance level on which statistical tests of adjusted quantities are carried. Actual type of reference standard deviation m_{0a} specifies whether during statistical analysis we use an a priori reference standard deviation m_0 or an a posteriori estimate m_0'. On the type of actual reference standard deviation depends the choice of density functions of stochastic quantities in statistical analysis of the adjustment.

A priori reference standard deviation m_0 is an estimate of the standard deviation of an observation with the unit weight. Numerically it is a scaling factor used in calculation of the weights. If we change m_0 , only the sum of weighted residuals squares is changed and all adjustment results remain the same (there is just one least squares solution). m_0 can be selected in cases when we know its value in advance and with sufficient reliability. Another situation when m_0 is used are networks with low number of degrees of freedom (poorly overdetermined systems) or when veen degrees of freedom is zero. Examples may be analysis of network models etc.

A posteri estimate of reference standard deviation m_0' is used in cases when a priori value of reference standard deviation m_0 is not known and when degrees of freedom is sufficiently high and reliable for empirical estimate of m_0' .

The standard deviation of an adjusted quantity θ is computed in dependece on the choice of actual type of reference standard deviation m_{0a} according to formula

$$m_{\theta_i} = m_{0a}\sqrt{q_{\theta_{ii}}},$$

where $q_{\theta_{ii}}$ is weight coefficient (cofactor) of the i-th adjusted unknown parameter (coordinate or orientation, $\theta = x_i$) or i-th adjusted observation (distance, direction, $\dots, \theta = L_i$).

Apart from standard deviation m_θ, program `gama-local` computes for adjusted quantity θ its *confidence interval* (Θ_1, Θ_2) in which the real value Θ is located with probability P

$$P(\Theta_1 < \Theta < \Theta_2) = P,$$

$$\Theta_1 = \theta - k_p m_\theta, \qquad \Theta_2 = \theta + k_p m_\theta,$$

where coefficient k_p depends on confidence probability P and in the case of low number of degrees of freedom on the choice of actual type of reference standard deviation m_{0a} .

Coefficient k_p is computed for $m_{0a} = m_0$ as critical value of normal distribution for probability $\alpha/2$, for the case of choice $m_{0a} = m_0'$ as critical value of Student distribution on confidence level $\alpha/2$ with τ degrees of freedom

$$k_p = \begin{cases} u_{\alpha/2} & \text{if } m_{0a} = m_0, \\ t_{\alpha/2,\tau} & \text{if } m_{0a} = m_0'. \end{cases}$$

Similarly confidence ellipses for adjusted points are defined in the following text.

4.4 Test on the reference standard deviation

Null hypothesis $H_0 : m_0 = m_0'$ is tested versus alternative hypothesis $H_1 : m_0 \neq m_0'$. Test criterion is ratio of a posteriori estimate of reference standard deviation

$$m_0' = \sqrt{\mathbf{v'Pv}/\tau}$$

and a priori reference standard deviation m_0 (input data parameter `m0-apr`). For given significance level α lower and upper bounds of interval (L, U) are computed so, that if hypothesis H_0 is true, probabilities $P(m_0'/m_0 \leq D)$ and $P(m_0'/m_0 \geq H)$ are equal to $\alpha/2$. Lower and upper bounds of the interval are computed as

$$L = \sqrt{(\chi_{1-\alpha/2,\tau}^2/\tau)}, \qquad U = \sqrt{(\chi_{\alpha/2,\tau}^2/\tau)}.$$

Probability

$$P(L < m_0'/m_0 < U) = \mathtt{conf - pr}$$

is by default 95%, this corresponds to 5% confidence level test.

Exceeding the upper limit H of the confidence interval can be caused even by a single gross error (one outlying observation). Method of Least Squares is generally very sensitive to presence of outliers. Safely can be detected only one observation whose elimination leads to maximal decrease of a posteriori estimate of reference standard deviation

$$m_0'' = \sqrt{(\mathbf{v'Pv} - \delta)/(\tau - 1)}, \qquad \delta = \max(v_i^2/q_{v_i}), \tag{6}$$

where

$$q_{v_i} = 1/p_i - q_{L_i} \tag{7}$$

is weight coefficient of i -th residual. If the set of observations contains only one gross error, the outlying observation is likely to be detected, but this can not be guaranteed.

In addition, program `gama-local` computes a posteriori estimate of reference standard deviation separately for horizontal distances and directions and/or angles after formula from

$$m_{0t}' = \sqrt{\sum \tilde{v}_{i_t}^2 / \sum \tilde{q}_{v_{i_t}}}, \qquad t = d, s,$$

where symbol t denotes observed distances, directions and/or angles.

Example

```
    m0  apriori  :     10.00
    m0' empirical:      9.64           [pvv] : 3.43560e+03


    During statistical analysis we work


    - with empirical standard deviation 9.64
    - with confidence level           95 %


    Ratio m0' empirical / m0 apriori: 0.964
    95 % interval (0.773, 1.227) contains value m0'/m0
```

```
m0'/m0 (distances): 0.997    m0'/m0 (directions): 0.943

Maximal decrease of m0''/m0 on elimination of one observation: 0.892

Maximal studentized residual 2.48 exceeds critical value 1.95
on significance level 5 % for observation #35
<distance from="407" to="422" val="346.415" stdev="5.0" />
```

4.5 Information on points

Program `gama-local` lists separately review of coordinates of fixed and adjusted points;
adjusted *constrained* coordinates are marked with *; see equation (3). Adjusted coordinate
standard deviations m_x and m_y , and values for computing confidence intervals are given in
the listing of adjusted coordinates (Section 4.3 [Statistical analysis], page 34). In the review
index i is the index of unknown x_i from the system of project equations (1) corresponding
to the point coordinates x and y .

Example

```
Fixed points
************

         point        x               y
     =========================================

         1   1054980.484      644498.590
         2   1054933.801      643654.101


Adjusted coordinates
********************

    i        point   approximate  correction  adjusted    std.dev conf.i.
=====================   value ====== [m] ====== value ========== [mm] ===

         422
    2            x   1055167.22747  -0.00510 1055167.22237     2.7     5.4
    3            y    644041.46119   0.00023  644041.46142     2.5     5.1

         424
    4            X *  1055205.41198  -0.00056 1055205.41142     3.1     6.3
    5            Y *   644318.24425  -0.00125  644318.24300     3.6     7.2
```

For adjusted points, program summarizes information on standard ellipses, confidence ellipses, mean square positional errors (m_p), mean coordinate errors (m_{xy}) and coefficients
g characterizing position of approximate coordinates with regard to the confidence ellipse.

Example

```
Mean errors and parameters of error ellipses
********************************************
```

point	mp [mm]	mxy [mm]	a [mm]	mean error ellipse b	alpha[g]	conf.err. ellipse a' [mm]	b'	g
422	3.6	2.6	2.7	2.5	187.0	6.8	6.4	0.8
424	4.7	3.4	3.7	2.9	131.8	9.5	7.4	0.2
403	5.7	4.0	4.3	3.6	78.9	11.0	9.3	1.1

Mean square positional error m_p and mean coordinate error (m_{xy}) are computed as

$$m_p = \sqrt{m_y^2 + m_x^2}, \qquad m_{xy} = m_p/\sqrt{2},$$

where m_y^2 and m_x^2 are squares of standard deviations (variances) of adjusted points coordinates.

Semimajor and semiminor axes of standard ellipse are denoted as a and b in the listing, bearing of semimajor axis is denoted as α and they are computed from covariances of adjusted coordinates

$$a = \sqrt{\frac{1}{2}(\text{cov } yy + \text{cov } xx + c)}, \qquad b = \sqrt{\frac{1}{2}(\text{cov } yy + \text{cov } xx - c)},$$

$$c = \sqrt{(\text{cov } xx - \text{cov } yy)^2 + 4(\text{cov } xy)^2},$$

$$\tan 2\alpha = 2(\text{cov } xy)/(\text{cov } xx - \text{cov } yy).$$

The angle α (the bearing of semimajor axis) is measured clockwise from X axis.

Probability that standard ellipse covers real position of a point is relatively low. For this reason program `gama-local` computes extra *confidence ellipse* for which the probability of covering real point position is equal to the given confidence probability. Both ellipsy are located in the same center, they share the same bearing of semimajor axes and they are similar. For lengths of their semi-axis holds

$$a' = k_p a, \qquad b' = k_p b,$$

where k_p is a coefficient computed for the given probability P as defined in Section 4.3 [Statistical analysis], page 34.

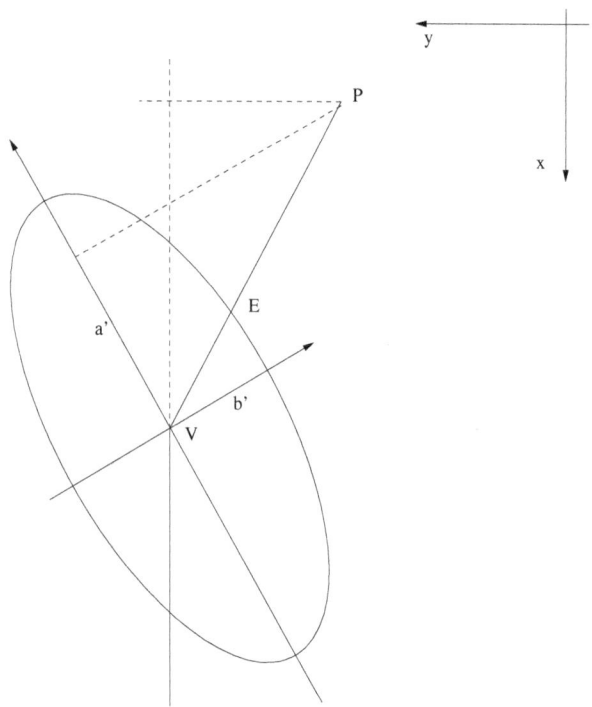

Position of approximate coordinates of an adjusted point with respect to its confidence ellipse is described by two points P and V where point P depicts approximate coordinates and V adjusted coordinates. Point E is the intersection of oriented half-line VP and the confidence ellipse. Coefficient g is defined as the ration of abscissae

$$g = \overline{VP}/\overline{VE}.$$

Three cases are possible

   $g < 1$ approximate coordinates of adjusted point are located inside the confidence ellipse

   $g = 1$ approximate coordinates of adjusted point are located on the confidence ellipse

   $g > 1$ approximate coordinates of adjusted point are outside the confidence ellipse

The coefficient g is calculated from formula

$$g = \sqrt{(a_0/a')^2 + (b_0/b')^2}$$

where

$$b_0 = \delta_y \cos \alpha - \delta_x \sin \alpha, \qquad a_0 = \delta_y \sin \alpha - \delta_x \cos \alpha$$

symbol δ is used for correction of approximate coordinates and α is bearing of confidence ellipse semimajor axis.

If network contains sets of observed directions, program writes information on corresponding adjusted orientations, standard deviations and confidence intervals. Index i is the same as in the case of adjusted coordinates the index of i-th adjusted unknown in the project equations.

Example

```
Adjusted bearings
*****************
```

i	standpoint	approximate value [g]	correction [g]	adjusted value [g]	std.dev	conf.i. [cc]
1	1	296.484371	-0.000917	296.483454	5.1	10.3
10	2	96.484371	0.000708	96.485079	5.1	10.4
21	403	20.850571	-0.001953	20.848618	8.8	17.7

4.6 Adjusted observations and residuals

In the review of adjusted observations program `gama-local` prints index of the observation, index of the row in matrix **A** in the system (1), identifications of standpoint and target point, type of the observation, its approximate and adjusted value, standard deviation and confidence interval.

Example

```
Adjusted observations
********************
```

i	standpoint	target		observed value	adjusted [m|g]	std.dev	conf.i. [mm|cc]
1	1	2	dis.	845.77700	845.77907	3.0	6.1
2		422	dir.	28.205700	28.205613	5.1	10.3
3		424	dir.	60.490600	60.491359	6.7	13.6

Review of residuals serves for analysis of observations and containts values of normalized or studentized residuals (depending on type of m_{0a} used) and three characteristics. Theese are coefficient **f** identifying weak network elements and estimates of real error of observation **e-obs** and real error of its adjusted value **e-adj**, see definition in the following text.

If normalized or studentized residual exceeds critical value for the given confidence probability, it is marked in the review with symbol **c** (critical) and maximal normalized or studentized residual is marked with symbol **m**.

Example

```
Residuals and analysis of observations
**************************************
```

i	standpoint	target		f[%]	v [mm|cc]	|v'|	e-obs. [mm|cc]	e-adj.
1	1	2	dir.	47.4	9.170	1.1	12.7	3.5

| 2 | 422 dir. 47.0 | -0.873 0.1 | -1.2 -0.3 |
| 3 | 424 dir. 30.3 | 7.588 1.1 | 14.8 7.2 |

4.6.1 Test on normal distribution of homogenized residuals

Repeated observations often display a normal frequency distribution. Residual of observed quantities are linear function real errors. From presumption of normal ditribution of real errors follows that homogenized residuals should have normal distribution as well.

Program `gama-local` estimates mean value $E(\tilde v)$ and estimate of variance $V(\tilde v)$ for the vector of homogenized residuals $$ E(\tilde v) = 1\over N\sum_i=1^N\tilde v_i, \qquad V(\tilde v) = E(\tilde v^2) - (E(\tilde v))^2. $$ Vector of homogenized residuals transforms to *normalized (standardized) vector of residuals* $$ \nu_i = \tilde v_i - E(\tilde v)\over\sqrt V(\tilde v). $$ Using Kolmogorov-Smirnov test program `gama-local` verifies assumption of normality of elements of vector $\bf \nu$. Result of the test is a value saying what is the probability that elements of vector $\bf \nu$ are a random sample form normal distribution $N(0, 1).$

Kolmogorov-Smirnov test for one sample is based on maximal difference between empirical and theoretical cumulative distribution function (normal distribution $N(0, 1)$ in our case). For random sample with N elements $X_1, X_2, \ldots, X_N$ from the population with cumulative distribution function $F(x)$ we form empirical cumulative distribution function $$ S_N(x) = (\hbox number of elements $X_1,X_2,\ldots,X_N$ which are \le x) / N, $$ %%% ****** as is schematically depicted on Fig. \refvyrovnani:obrks. If we denote $$ D = \max_-\infty < x < \infty|S_N(x) - F(x)|, $$ the testing criterion $D\sqrt N$ has limit of Kolmogorov-Smirnov distribution. Some critical values of testing criterion $D\sqrt N$ computed from the KS distribution are given in the following table $$ \vbox\offinterlineskip \halign \strut \vrule\quad # \quad &\vrule \hfil\quad # \quad\hfil & \hfil\quad # \quad\hfil & \hfil\quad # \quad\hfil & \hfil\quad # \quad\hfil & \hfil\quad # \quad\hfil\vrule \cr \noalign\hrule & 0.005 & 0.010 & 0.025 & 0.050 & 0.100 \cr \noalign\hrule Lower & 0.42 & 0.44 & 0.48 & 0.52 & 0.57 \cr Upper & 1.73 & 1.63 & 1.48 & 1.36 & 1.22 \cr \noalign\hrule $$

4.7 Identification of weak network elements

When planning observations in a geodetic network we always try to guarantee that all observed elements are checked by other measurements. Only with redundant measurements it is possible to adjust observations and possibly remove blunders that might otherwise totaly corrupt the whole set of measurements. Apart from sufficient number of redundant observations the degree of control of single observed elements is given by the network configuration, ie. its geometry.

Less controlled observations represent weak network elements and they can in extreme cases even disable detection of gross observational errors as it is in the case of uncontrolled observations. There are two limit cases of observation control

> **fully controlled observation** as is for example an observed distance between two fixed points (standard deviation of the adjusted element is zero; standard deviation of the residual equals to the standard deviation if the observation) and

uncontrolled observations as is a free polar bar for example (standard deviation of adjusted value is equal to standard deviation of observed quantity; residual and standard deviation of the residual are zero).

Weakly controlled or uncontrolled observations can result even from elimination of certain suspisios observations during analysis of adjusment.

Standard deviation of adjusted observations is less than standard deviation of the measurement. Degree of observation control in network is defined as coefficient

$$f = 100\frac{m_\ell - m_L}{m_\ell}, \tag{8}$$

where m_ℓ is standard deviation of observed quantity and m_L is standard deviation computed from a posteriori reference standard deviation m_0. We consider observed network element to be

uncontrolled if $f < 0.1$ (in listing marked with letter `u`),

weakly controlled if $0.1 \le f < 5$ (in listing marked with letter `w`).

4.8 Estimation of real errors

Acording to previous section we can consider an observation to be controlled if its coefficient $f > 0.1$. Any controlled observation can be eliminated from the network without corrupting the network consistency—network reduced by one controlled observation can be adjusted and all unknown parameters can be compute without the eliminated observation.

Estimate of real error of i-th observation is defined as

$$\varepsilon_{\ell_i} = L_i^{red} - \ell_i, \tag{9}$$

where ℓ_i is value of i-th observation and L_i^{red} is value of i-th network element computed from adjusted coordinates and/or orientations of the reduced network. Similarly is defined the estimate of real error of a residual

$$\varepsilon_{v_i} = L_i^{red} - L_i. \tag{10}$$

Adjustment results are the best statistical estimate of unknown parameters that we have. This holds true even for adjustment of *reduced* network which is not influenced by real error of i-th observation. On favourable occasions differences (9) and (10) can help to detect blunders but to interpret these estimates as *real errors* is possible only with substantial exaggeration. These estimates fail when there are more than one significant observational error. Generally holds tha the weaker the element is controlled in netowrk the less reliable these estimates are.

Estimate of real error of an observation computes program `gama-local` as

$$\varepsilon_{\ell_i} = v_i/(p_i q_{v_i})$$

and estimate of real error of a residual as

$$\varepsilon_{v_i} = \varepsilon_{l_i} - v_i.$$

4.9 Test on linearization

Mathematical model of geodetic network adjustment in `gama-local` is defined as a set of known real-valued differentiable functions

$$\mathbf{L}^* = \varphi(\mathbf{X}^*), \tag{11}$$

where $\mathbf{L}^*$ is a vector of theoretical correct observations and $\mathbf{X}^*$ is a vector of correct values of parameters. For the given sample set of observations $\mathbf{L}$ and the unknown vector of residuals $\mathbf{v}$ we can express the estimate of parameters $\mathbf{X}$ as a nonlinear set of equations

$$\mathbf{L} + \mathbf{v} = \varphi(\mathbf{X}). \tag{12}$$

With approximate values $\mathbf{X_0}$ of unknown parameters

$$\mathbf{X} = \mathbf{X_0} + \mathbf{x}$$

we can linearize the equations (12)

$$\mathbf{L} + \mathbf{v} = \varphi(\mathbf{X_0}) + \left.\frac{\partial \varphi}{\partial \mathbf{X}}\right|_{\mathbf{X}=\mathbf{X_0}} \mathbf{x}$$

yielding the linear set of equations (1) where

$$\mathbf{A} = \left.\frac{\partial \varphi}{\partial \mathbf{X}}\right|_{\mathbf{X}=\mathbf{X_0}} \quad \text{and} \quad \mathbf{b} = \mathbf{L} - \varphi(\mathbf{X_0}).$$

Unknown parameters in `gama-local` mathematical model are points coordinates and orientation angles (transforming observed directions to bearings). The observables described by functions (12) belong into two classes

linear observables: horizontal and slope distances, height differences, control coordinates and vectors (coordinate differences),

angular observables: directions, horizontal and zenith angles.

Internally in `gama-local` unknown corrections to linear observables are computed in millimeters and corrections to angular observables in centigrade seconds. To reflect the internal units in used all partial derivatives of angular observables by coordinates are scaled by factor $2000/\pi = 10^{-3} \times (200 \times 10^4/\pi)$.

When computing coefficients of project equations (1) we expect that approximate coordinates of points are known with sufficient accuracy needed for linearization of generally nonlinear relations between observations and unknown paramters. Most often this is true but not always and generally we have to check how close our approximation is to adjusted parameters.

Generally we check linearization in adjustment by double calculation of residuals

$$\mathbf{v}^i = \mathbf{A}\mathbf{x} - \mathbf{b},$$
$$\mathbf{v}^{ii} = \bar{\ell}(\bar{\mathbf{x}}) - \ell,$$

where in our notation $\mathbf{x}$ is vector of corrections of approximate unknown parameters $\mathbf{x_0}$, $\mathbf{b}$ vector of reduced observations, ℓ vector of observations and $\bar{\ell}(\bar{\mathbf{x}})$ is vector of adjusted observations conputed from adjusted coordiantes $\bar{\mathbf{x}} = \mathbf{x_0} + \mathbf{x}$. Disagreement $\mathbf{v}^i \neq \mathbf{v}^{ii}$ signals discrepancies in linearization.

Program `gama-local` similarly computes and tests differences in values of adjusted observations once computed from residuals and once from adjusted coordinates. For measured directions and angles `gama-local` computes in addition transverse deviation corresponding to computed angle difference in the distance of target point (or the farther of two targets for angle). As a criterion of bad linearization is supposed positional deviation greater or equal to 0.0005 millimetres.

Example

```
Test of linearization error
***************************


Diffs in adj. obs from residuals and from adjusted coordinates
**************************************************************

  i standpoint     target        observed      r        difference
================================   value   = [mm|cc] = [cc] == [mm]=


  2 3022184030 3022724008 dist.    28.39200   -7.070            -0.003
  3            3022724002 dist.    72.30700  -18.815            -0.001
  7            3000001063 dir.    286.305200  11.272  -0.002    -0.001
  8            3022724008 dir.    357.800600 -23.947   0.037     0.002
```

From the practical point of view it might seem that the tolerance 0.0005 mm for detecting poor linearization is too strict. Its exceeding in program `gama-local` results in repeated adjustment with substitute adjusted coordinates for approximate. Given tolerance was chosen so strict to guarantee that listed output results would never be influenced by linearization and could serve for verification and testing of numerical solutions produced by other programs.

Implicitly coordinates of constrained points are not changed in iterative adjustments. This feature can be changed in XML input data by setting `<parameters update-constrained-coordinates = "yes" />` (see Section 2.5 [Network parameters], page 9).

Iterated adjustement with successive improvement of approximate coordinates converges usually even for gross errors in initial estimates of unknown coordinates. If the influence of linearization is detected after adjustment, quite often only one iteration is sufficient for recovering.

For any automatically controlled iteration we have to set up certain stopping criterion independent on the convergence and results obtained. Program `gama-local` computes iterated adjustment three times at maximum. If the bad linearization is detected even after three readjustments it signals that given network configuration is somehow suspicious.

5 Data structures and algorithms

5.1 Observation data and points

The Gama observation data structures are designed to enable adjustment of any combination of possibly correlated observations. At its very early stage Gama was limited to adjustment of uncorrelated observations. Only directions and distances were available and observable's weight was stored together with the observed value in a single object. A single array of pointers to observation objects was sufficient for handling all observations. So called *orientation shifts* corresponding to directions measured form a point were stored together with coordinations in *point objects*.

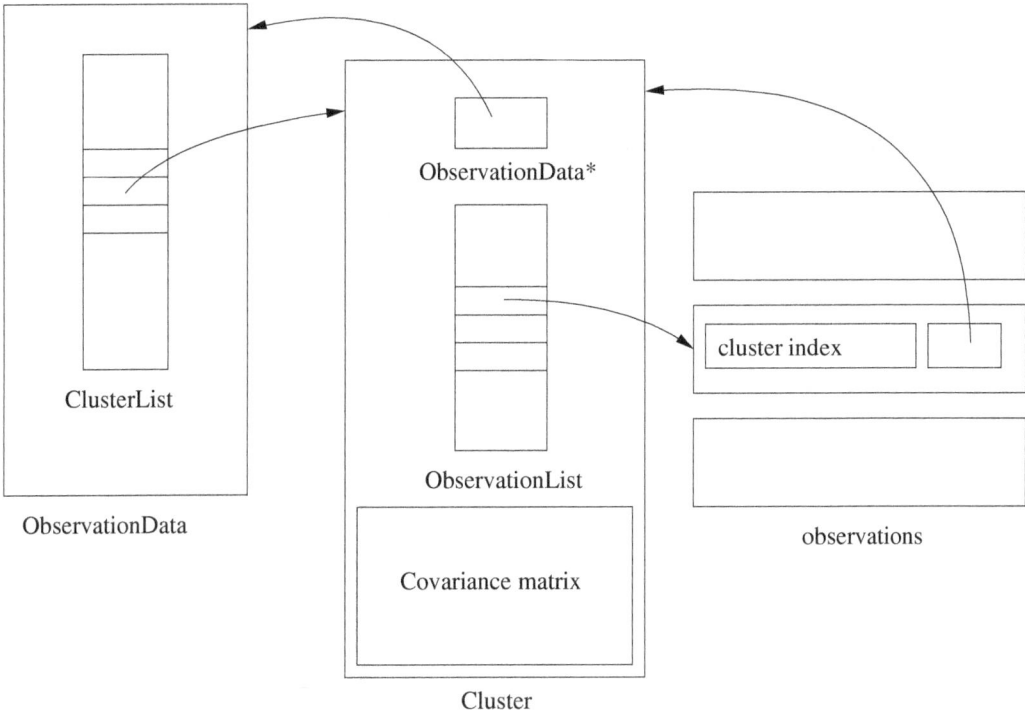

To enable adjustment of possibly correlated observations (like angles derived from observed directions or already adjusted coordinates from a previous adjustment) Gama has come with the concept of *clusters*. Cluster is an object with a common variance-covariance matrix and a list of pointers to observation objects (distances, directions, angles, etc.). Weights were removed from observation objects and replaced with a pointer to the cluster to which the observation belong. All clusters are joined in a common object ObservationData; similarly to observations, each cluster contains a pointer to its parent Observation Data object. *Orientation shifts* were separated from coordinates and are stored in the cluster containing the bunch of directions and thus number of orientations is not limited to one for a point.

This organisation of observational information has proved to be effective. Template classes ObservationData and Cluster are used as base classes both in gama-local and gama-g3

```
template <typename Observation>
  class ObservationData
```

```
{
public:
  ClusterList<Observation>  CL;

  ObservationData();
  ObservationData(const ObservationData& cod);
  ~ObservationData();

  ObservationData& operator=(const ObservationData& cod);
  template <typename P> void for_each(const P& p) const;
};

template <typename Observation>
  class Cluster
  {
  public:
    const ObservationData<Observation>*    observation_data;
    ObservationList<Observation>           observation_list;
    typename Observation::CovarianceMatrix covariance_matrix;

    Cluster(const ObservationData<Observation>* od);
    virtual ~Cluster();

    virtual Cluster* clone(const ObservationData<Observation>*) const = 0;
    double stdDev(int i) const;
    int size() const;
    void update();
    int  activeCount() const;
    typename Observation::CovarianceMatrix activeCov() const;
  };
```

The following template class `PointBase` for handling point information is used in **gama-g3**. The template class `PointBase` relies internally on `std::map` container but comes with its own interface (in **gama-local** `std::map` was used directly for storing points).

```
template <typename Point>
  class PointBase
  {
    typedef std::map<typename Point::Name, Point*>  Points;

  public:
    PointBase();
    PointBase(const PointBase& cod);
    ~PointBase();

    PointBase& operator=(const PointBase& cod);
    void put(const Point&);
```

```
    void put(Point*);
    Point*      find(const typename Point::Name&);
    const Point* find(const typename Point::Name&) const;
    void erase(const typename Point::Name&);
    void erase();

    class const_iterator;
    const_iterator  begin();
    const_iterator  end  ();

    class iterator;
    iterator  begin();
    iterator  end  ();
  };
```

Template classes `ObservationData` and `PointBase` are defined in namespace `GNU_gama` and are located in the source directory `gnu_gama`.

5.2 Supported ellipsoids

id	a	b, 1/f, f	description	
airy	6377563.396	6356256.910	Airy ellipsoid 1830	[4]
airy_mod	6377340.189	6356034.446	Modified Airy	[4]
apl1965	6378137	298.25	Appl. Physics. 1965	[4]
andrae1876	6377104.43	300.0	Andrae 1876 (Denmark, Iceland)	[4]
australian	6378160	298.25	Australian National 1965	[3]
bessel	6377397.15508	6356078.96290	Bessel ellipsoid 1841	[1]
bessel_nam	6377483.865	299.1528128	Bessel 1841 (Namibia)	[4]
clarke1858a	6378361	6356685	Clarke ellipsoid 1858 1st	[3]
clarke1858b	6378558	6355810	Clarke ellipsoid 1858 2nd	[3]
clarke1866	6378206.4	6356583.8	Clarke ellipsoid 1866	[3]
clarke1880	6378316	6356582	Clarke ellipsoid 1880	[3]
clarke1880m	6378249.145	293.4663	Clarke ellipsoid 1880 (modified)	[4]
cpm1799	6375738.7	334.29	Comm. des Poids et Mesures 1799	[4]
delambre	6376428	311.5	Delambre 1810 (Belgium)	[4]
engelis	6378136.05	298.2566	Engelis 1985	[4]
everest1830	6377276.345	300.8017	Everest 1830	[4]
everest1848	6377304.063	300.8017	Everest 1948	[4]
everest1856	6377301.243	300.8017	Everest 1956	[4]
everest1869	6377295.664	300.8017	Everest 1969	[4]

everest_ss	6377298.556	300.8017	Everest (Sabah and Sarawak)	[4]
fisher1960	6378166	298.3	Fisher 1960 (Mercury Datum)	[3] [4]
fisher1960m	6378155	298.3	Modified Fisher 1960	[3] [4]
fischer1968	6378150	298.3	Fischer 1968	[4]
grs67	6378160	298.2471674270	GRS 67 (IUGG 1967)	[4]
grs80	6378137	298.257222101	Geodetic Reference System 1980	[1]
hayford	6378388	297	Hayford 1909 (International)	[1] [3]
helmert	6378200	298.3	Helmert ellipsoid 1906	[3]
hough	6378270	297	Hough	[4]
iau76	6378140	298.257	IAU 1976	[4]
international	6378388	297	International 1924 (Hayford 1909)	[1] [3]
kaula	6378163	298.24	Kaula 1961	[4]
krassovski	6378245	298.3	Krassovski ellipsoid 1940	[1]
lerch	6378139	298.257	Lerch 1979	[4]
mprts	6397300	191.0	Maupertius 1738	[4]
mercury	6378166	298.3	Mercury spheroid 1960	[3]
merit	6378137	298.257	MERIT 1983	[4]
new_intl	6378157.5	6356772.2	New International 1967	[4]
nwl1965	6378145	298.25	Naval Weapons Lab., 1965	[4]
plessis	6376523	6355863	Plessis 1817 (France)	[4]
se_asia	6378155	6356773.3205	Southeast Asia	[4]
sgs85	6378136	298.257	Soviet Geodetic System 85	[4]
schott	6378157	304.5	Schott 1900 spheroid	[3]
sa1969	6378160	298.25	South American Spheroid 1969	[3]
walbeck	6376896	6355834.8467	Walbeck	[4]
wgs60	6378165	298.3	WGS 60	[4]
wgs66	6378145	298.25	WGS 66	[4]
wgs72	6378135	298.26	WGS 72	[4]
wgs84	6378137	298.257223563	World Geodetic System 1984	[1]

[1]   Milos Cimbalnik - Leos Mervart: Vyssi geodezie 1, 1997, Vydavatelstvi CVUT, Praha

[2]   Milos Cimbalnik: Derived Geometrical Constants of the Geodetic Reference System 1980, Studia geoph. et geod. 35 (1991), pp. 133-144, NCSAV, Praha

[3]   Glossary of the Mapping Sciences, Prepared by a Joint Committe of the American Society of Civil Engineers, American Congress on Surveying and Mapping and American Society for Photogrammetry and Remote Sensing (1994), USA, ISBN 1-57083-011-8, ISBN 0-7844-0050-4

[4]   Gerald Evenden: proj - forward cartographic projection filter (rel.   4.3.3), http://www.remotesensing.org/proj

5.3 Transformation from spatial to geographical coordinates

Spatial coordinates (X, Y, Z) can be easily computed from geographical ellipsoidal coordinates (B, L, H), where B is geographical latitude, L geographical longitude and H is elliposidal height, as

$$\begin{pmatrix} X \\ Y \\ Z \end{pmatrix} = \begin{pmatrix} (N + H) \cos B \cos L \\ (N + H) \cos B \sin L \\ (N(1 - e^2) + H) \sin B \end{pmatrix}$$

where $N = a/\sqrt{1 - e^2 \sin^2 B}$ is the radius of curvature in the prime vertical, $e^2 = (a^2 - b^2)/a^2$ is the first eccentricity for the given rotational ellipsoid (spheroid) with semi-major axis a and semi-minor axis b.

In the case of coordiante transformation from (X, Y, Z) to (B, L, H), the longitude is given by the formula

$$\tan L = Y/X.$$

Now we can introduce $D = \sqrt{X^2 + Y^2}$, so that the cartesian system become (D, Z). Coordinates B and H are then usually computed by iteration with some starting value of B_0, for example $\tan B_0 = Z/D/(1 - e^2)$,

$$\tan B_i = Z/D + \frac{N_{i-1}}{(N_{i-1} + H_{i-1})} e^2 \tan B_{i-1}, \quad H_i = D/\cos B_{i-1} = Z/\sin B_{i-1} - N(1 - e^2)$$

B. R. Bowring described a closed formula[1] that is more effective and sufficiantly accurate and that is used in GNU Gama.

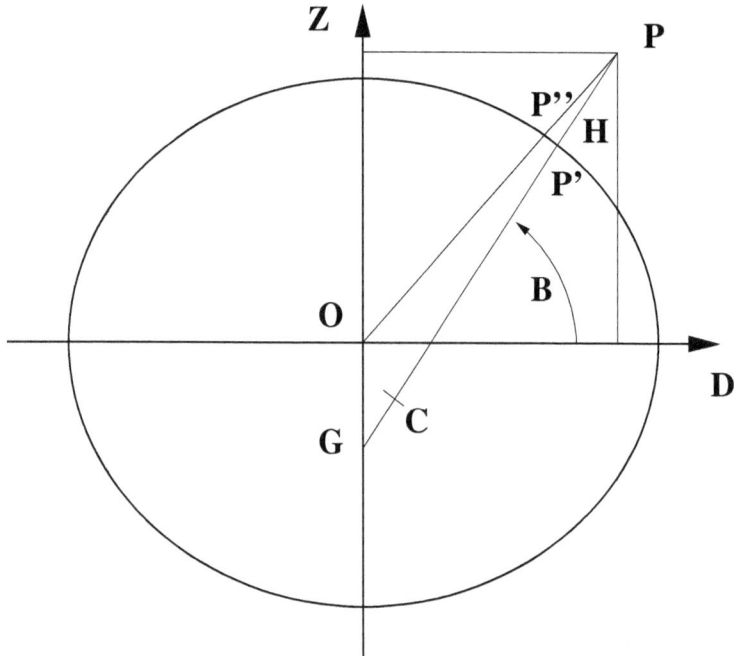

The centre of curvature C of the spheroid corresponding to P' is the point

$$(e^2 a \cos^2 u, -e'^2 b \sin^3 u),$$

where $e'^2 = (a^2 - b^2)/b^2$ is second eccentricity and u is the parametric latitude of the point P', $(1 - e^2)N \sin B = b \sin u$. Therefore

$$\tan B = \frac{Z + e'^2 b \sin^3 u}{D - e^2 a \cos^3 u}.$$

This is clearly an iterative solution; but it has been found that this formula is extremely accurate using the single first approximation for u for the $\tan u = (Z/D)(a/b)$. Maximum error in earth bound region is 3e-8 of sexadecimal arc seconds (5e-7 millimetres); maximum is 0.0018" (0.1 millimetres) at height H = 2a.

5.4 Class g3::Model

g3::model documentation shall come here ...

```
namespace GNU_gama {  namespace g3 {


  class Model {
```

---

[1] B. R. Bowring: Transformation from spatial to geographical coordinates, Survey Review XXIII, 181, July 1976

```
  public:

    typedef GNU_gama::PointBase<g3::Point>              PointBase;
    typedef GNU_gama::ObservationData<g3::Observation>  ObservationData;

    PointBase            *points;
    ObservationData      *obs;

    GNU_gama::Ellipsoid  ellipsoid;


    Model();
    ~Model();

    Point* get_point(const Point::Name&);
    void   write_xml(std::ostream& out) const;
    void   pre_linearization();
}}
```

Appendix A Copying This Manual

A.1 GNU Free Documentation License

Version 1.1, March 2000

Copyright © 2000 Free Software Foundation, Inc.
51 Franklin Street, Fifth Floor, Boston, MA 02110-1301, USA

Everyone is permitted to copy and distribute verbatim copies
of this license document, but changing it is not allowed.

0. PREAMBLE

The purpose of this License is to make a manual, textbook, or other written document *free* in the sense of freedom: to assure everyone the effective freedom to copy and redistribute it, with or without modifying it, either commercially or noncommercially. Secondarily, this License preserves for the author and publisher a way to get credit for their work, while not being considered responsible for modifications made by others.

This License is a kind of "copyleft", which means that derivative works of the document must themselves be free in the same sense. It complements the GNU General Public License, which is a copyleft license designed for free software.

We have designed this License in order to use it for manuals for free software, because free software needs free documentation: a free program should come with manuals providing the same freedoms that the software does. But this License is not limited to software manuals; it can be used for any textual work, regardless of subject matter or whether it is published as a printed book. We recommend this License principally for works whose purpose is instruction or reference.

1. APPLICABILITY AND DEFINITIONS

This License applies to any manual or other work that contains a notice placed by the copyright holder saying it can be distributed under the terms of this License. The "Document", below, refers to any such manual or work. Any member of the public is a licensee, and is addressed as "you".

A "Modified Version" of the Document means any work containing the Document or a portion of it, either copied verbatim, or with modifications and/or translated into another language.

A "Secondary Section" is a named appendix or a front-matter section of the Document that deals exclusively with the relationship of the publishers or authors of the Document to the Document's overall subject (or to related matters) and contains nothing that could fall directly within that overall subject. (For example, if the Document is in part a textbook of mathematics, a Secondary Section may not explain any mathematics.) The relationship could be a matter of historical connection with the subject or with related matters, or of legal, commercial, philosophical, ethical or political position regarding them.

The "Invariant Sections" are certain Secondary Sections whose titles are designated, as being those of Invariant Sections, in the notice that says that the Document is released under this License.

The "Cover Texts" are certain short passages of text that are listed, as Front-Cover Texts or Back-Cover Texts, in the notice that says that the Document is released under this License.

A "Transparent" copy of the Document means a machine-readable copy, represented in a format whose specification is available to the general public, whose contents can be viewed and edited directly and straightforwardly with generic text editors or (for images composed of pixels) generic paint programs or (for drawings) some widely available drawing editor, and that is suitable for input to text formatters or for automatic translation to a variety of formats suitable for input to text formatters. A copy made in an otherwise Transparent file format whose markup has been designed to thwart or discourage subsequent modification by readers is not Transparent. A copy that is not "Transparent" is called "Opaque".

Examples of suitable formats for Transparent copies include plain ASCII without markup, Texinfo input format, LaTeX input format, SGML or XML using a publicly available DTD, and standard-conforming simple HTML designed for human modification. Opaque formats include PostScript, PDF, proprietary formats that can be read and edited only by proprietary word processors, SGML or XML for which the DTD and/or processing tools are not generally available, and the machine-generated HTML produced by some word processors for output purposes only.

The "Title Page" means, for a printed book, the title page itself, plus such following pages as are needed to hold, legibly, the material this License requires to appear in the title page. For works in formats which do not have any title page as such, "Title Page" means the text near the most prominent appearance of the work's title, preceding the beginning of the body of the text.

2. VERBATIM COPYING

You may copy and distribute the Document in any medium, either commercially or noncommercially, provided that this License, the copyright notices, and the license notice saying this License applies to the Document are reproduced in all copies, and that you add no other conditions whatsoever to those of this License. You may not use technical measures to obstruct or control the reading or further copying of the copies you make or distribute. However, you may accept compensation in exchange for copies. If you distribute a large enough number of copies you must also follow the conditions in section 3.

You may also lend copies, under the same conditions stated above, and you may publicly display copies.

3. COPYING IN QUANTITY

If you publish printed copies of the Document numbering more than 100, and the Document's license notice requires Cover Texts, you must enclose the copies in covers that carry, clearly and legibly, all these Cover Texts: Front-Cover Texts on the front cover, and Back-Cover Texts on the back cover. Both covers must also clearly and legibly identify you as the publisher of these copies. The front cover must present the full title with all words of the title equally prominent and visible. You may add other material on the covers in addition. Copying with changes limited to the covers, as long as they preserve the title of the Document and satisfy these conditions, can be treated as verbatim copying in other respects.

If the required texts for either cover are too voluminous to fit legibly, you should put the first ones listed (as many as fit reasonably) on the actual cover, and continue the rest onto adjacent pages.

If you publish or distribute Opaque copies of the Document numbering more than 100, you must either include a machine-readable Transparent copy along with each Opaque copy, or state in or with each Opaque copy a publicly-accessible computer-network location containing a complete Transparent copy of the Document, free of added material, which the general network-using public has access to download anonymously at no charge using public-standard network protocols. If you use the latter option, you must take reasonably prudent steps, when you begin distribution of Opaque copies in quantity, to ensure that this Transparent copy will remain thus accessible at the stated location until at least one year after the last time you distribute an Opaque copy (directly or through your agents or retailers) of that edition to the public.

It is requested, but not required, that you contact the authors of the Document well before redistributing any large number of copies, to give them a chance to provide you with an updated version of the Document.

4. MODIFICATIONS

You may copy and distribute a Modified Version of the Document under the conditions of sections 2 and 3 above, provided that you release the Modified Version under precisely this License, with the Modified Version filling the role of the Document, thus licensing distribution and modification of the Modified Version to whoever possesses a copy of it. In addition, you must do these things in the Modified Version:

A. Use in the Title Page (and on the covers, if any) a title distinct from that of the Document, and from those of previous versions (which should, if there were any, be listed in the History section of the Document). You may use the same title as a previous version if the original publisher of that version gives permission.

B. List on the Title Page, as authors, one or more persons or entities responsible for authorship of the modifications in the Modified Version, together with at least five of the principal authors of the Document (all of its principal authors, if it has less than five).

C. State on the Title page the name of the publisher of the Modified Version, as the publisher.

D. Preserve all the copyright notices of the Document.

E. Add an appropriate copyright notice for your modifications adjacent to the other copyright notices.

F. Include, immediately after the copyright notices, a license notice giving the public permission to use the Modified Version under the terms of this License, in the form shown in the Addendum below.

G. Preserve in that license notice the full lists of Invariant Sections and required Cover Texts given in the Document's license notice.

H. Include an unaltered copy of this License.

I. Preserve the section entitled "History", and its title, and add to it an item stating at least the title, year, new authors, and publisher of the Modified Version as given on the Title Page. If there is no section entitled "History" in the Document, create one stating the title, year, authors, and publisher of the Document as given on its Title Page, then add an item describing the Modified Version as stated in the previous sentence.

J. Preserve the network location, if any, given in the Document for public access to a Transparent copy of the Document, and likewise the network locations given in the Document for previous versions it was based on. These may be placed in the "History" section. You may omit a network location for a work that was published at least four years before the Document itself, or if the original publisher of the version it refers to gives permission.

K. In any section entitled "Acknowledgments" or "Dedications", preserve the section's title, and preserve in the section all the substance and tone of each of the contributor acknowledgments and/or dedications given therein.

L. Preserve all the Invariant Sections of the Document, unaltered in their text and in their titles. Section numbers or the equivalent are not considered part of the section titles.

M. Delete any section entitled "Endorsements". Such a section may not be included in the Modified Version.

N. Do not retitle any existing section as "Endorsements" or to conflict in title with any Invariant Section.

If the Modified Version includes new front-matter sections or appendices that qualify as Secondary Sections and contain no material copied from the Document, you may at your option designate some or all of these sections as invariant. To do this, add their titles to the list of Invariant Sections in the Modified Version's license notice. These titles must be distinct from any other section titles.

You may add a section entitled "Endorsements", provided it contains nothing but endorsements of your Modified Version by various parties—for example, statements of peer review or that the text has been approved by an organization as the authoritative definition of a standard.

You may add a passage of up to five words as a Front-Cover Text, and a passage of up to 25 words as a Back-Cover Text, to the end of the list of Cover Texts in the Modified Version. Only one passage of Front-Cover Text and one of Back-Cover Text may be added by (or through arrangements made by) any one entity. If the Document already includes a cover text for the same cover, previously added by you or by arrangement made by the same entity you are acting on behalf of, you may not add another; but you may replace the old one, on explicit permission from the previous publisher that added the old one.

The author(s) and publisher(s) of the Document do not by this License give permission to use their names for publicity for or to assert or imply endorsement of any Modified Version.

5. COMBINING DOCUMENTS

You may combine the Document with other documents released under this License, under the terms defined in section 4 above for modified versions, provided that you include in the combination all of the Invariant Sections of all of the original documents, unmodified, and list them all as Invariant Sections of your combined work in its license notice.

The combined work need only contain one copy of this License, and multiple identical Invariant Sections may be replaced with a single copy. If there are multiple Invariant Sections with the same name but different contents, make the title of each such section unique by adding at the end of it, in parentheses, the name of the original author or publisher of that section if known, or else a unique number. Make the same adjustment to the section titles in the list of Invariant Sections in the license notice of the combined work.

In the combination, you must combine any sections entitled "History" in the various original documents, forming one section entitled "History"; likewise combine any sections entitled "Acknowledgments", and any sections entitled "Dedications". You must delete all sections entitled "Endorsements."

6. COLLECTIONS OF DOCUMENTS

You may make a collection consisting of the Document and other documents released under this License, and replace the individual copies of this License in the various documents with a single copy that is included in the collection, provided that you follow the rules of this License for verbatim copying of each of the documents in all other respects.

You may extract a single document from such a collection, and distribute it individually under this License, provided you insert a copy of this License into the extracted document, and follow this License in all other respects regarding verbatim copying of that document.

7. AGGREGATION WITH INDEPENDENT WORKS

A compilation of the Document or its derivatives with other separate and independent documents or works, in or on a volume of a storage or distribution medium, does not as a whole count as a Modified Version of the Document, provided no compilation copyright is claimed for the compilation. Such a compilation is called an "aggregate", and this License does not apply to the other self-contained works thus compiled with the Document, on account of their being thus compiled, if they are not themselves derivative works of the Document.

If the Cover Text requirement of section 3 is applicable to these copies of the Document, then if the Document is less than one quarter of the entire aggregate, the Document's Cover Texts may be placed on covers that surround only the Document within the aggregate. Otherwise they must appear on covers around the whole aggregate.

8. TRANSLATION

Translation is considered a kind of modification, so you may distribute translations of the Document under the terms of section 4. Replacing Invariant Sections with translations requires special permission from their copyright holders, but you may include translations of some or all Invariant Sections in addition to the original versions of these Invariant Sections. You may include a translation of this License provided that you also include the original English version of this License. In case of a disagreement between the translation and the original English version of this License, the original English version will prevail.

9. TERMINATION

You may not copy, modify, sublicense, or distribute the Document except as expressly provided for under this License. Any other attempt to copy, modify, sublicense or distribute the Document is void, and will automatically terminate your rights under this License. However, parties who have received copies, or rights, from you under this License will not have their licenses terminated so long as such parties remain in full compliance.

10. FUTURE REVISIONS OF THIS LICENSE

The Free Software Foundation may publish new, revised versions of the GNU Free Documentation License from time to time. Such new versions will be similar in spirit to the present version, but may differ in detail to address new problems or concerns. See http://www.gnu.org/copyleft/.

Each version of the License is given a distinguishing version number. If the Document specifies that a particular numbered version of this License "or any later version" applies to it, you have the option of following the terms and conditions either of that specified version or of any later version that has been published (not as a draft) by the Free Software Foundation. If the Document does not specify a version number of this License, you may choose any version ever published (not as a draft) by the Free Software Foundation.

A.1.1 ADDENDUM: How to use this License for your documents

To use this License in a document you have written, include a copy of the License in the document and put the following copyright and license notices just after the title page:

```
Copyright (C)  year  your name.
Permission is granted to copy, distribute and/or modify this document
under the terms of the GNU Free Documentation License, Version 1.1
or any later version published by the Free Software Foundation;
with the Invariant Sections being list their titles, with the
Front-Cover Texts being list, and with the Back-Cover Texts being list.
A copy of the license is included in the section entitled ``GNU
Free Documentation License''.
```

If you have no Invariant Sections, write "with no Invariant Sections" instead of saying which ones are invariant. If you have no Front-Cover Texts, write "no Front-Cover Texts" instead of "Front-Cover Texts being *list*"; likewise for Back-Cover Texts.

If your document contains nontrivial examples of program code, we recommend releasing these examples in parallel under your choice of free software license, such as the GNU General Public License, to permit their use in free software.

Concept Index